M. RIAZIAT

THE AGE
OF SIMORGH

planning for human convergence
with technology

California Scientific, Inc.

Cover design by Cindy Cheung

Interior design by Aaxel Author Services

ISBN (Paperback): 979-8-9875153-0-3
ISBN (Hardback): 979-8-9875153-1-0
ISBN (eBook): 979-8-9875153-2-7

Printed in the United States of America

Contents

Machine-assisted collective decision making that minimizes human supervision, signals the dawn of the age of Simorgh: the age of humans merging with their technology to create larger and more complex entities. We are currently observing the emergence of an early Simorgh who is learning to make decisions on its own using a combination of humans and machines. What are the possible outcomes, and what are the choices for individual humans?

Introduction

We are entering a new era of existence, one in which mankind will merge with technology to create a new way of life. Our form of government will become obsolete; egalitarianism will no longer be a lofty goal; the distinction between reality and fantasy will fade, and the lifestyle of humanity will change drastically to one of happiness and leisure.

Many common predictions and speculations about recent technological advancements and their effects on human lifestyle are dystopian or apocalyptic. Particularly, in the technical fields of robotics and artificial intelligence, popular fiction tends to show robots assuming a dominant position, forcing humans to engage in a struggle to maintain some control over the autonomous and intelligent machines.[1] Perceiving anything that disrupts our routine existence as a threat is a common reaction but it is not universal. Some authors argue that technology has always helped us to achieve a better standard of living, and that it will continue to do so. We dispute the universality of this viewpoint. Other authors have predicted the

---

[1] This is also seen in popular movie productions such as *The Terminator*, Orion Pictures 1984.

evolution of a "global brain" or a similar collective union[2]. The reader may deduce from these writings that humans are on the verge of losing their individuality. Conversely, the inferred message may be one of hope and excitement for a new evolutionary milestone.

In this book, we investigate the effect of technology on our society, our way of life, and our form of government. We are increasingly receiving collective or automated assistance in our daily decision making, and soon we will not be able to make unaided decisions on our own. Such signs and trends point in the direction of the evolution of a new type of organism with an immense potential for either advancement or failure. This entity will be in the form of a titanic, living, thinking, and acting organism whose body cells are humans and machines. One or multiple such organisms may emerge, but they will not engulf the entire human population. We call this new type of organism a SIMORGH (Sovereign IntelliMatic ORGanization of Humans).

In ancient Persian legends, Simorgh was depicted as a wise, benevolent, and magnificent bird with magical powers, who lived on a remote and inaccessible mountain top (Figure I-1).[3] One episode in "The Epic of Kings" describes how Simorgh adopts an albino child who has been abandoned by his parents, and raises him to prominence in the society as a distinguished warrior. In another more philosophical narration (The Conference of the Birds), a nation of birds embarks on a long expedition to seek the exalted Simorgh. The few who persevere the strenuous journey find themselves to have been collectively transmuted into the legendary Simorgh (Figure I-2).[4] See Appendix A for more detail.

---

[2] The evolution of a global organism was predicted as early as 1953 by Fred Kohler in "Evolution and Human Destiny." Science fiction writer Isaac Asimov imagined an entity named "Galaxia" that brought together the entire humanity in a super-organism with a collective mind. One of the more recent analyses was written by Gregory Stock in "Metaman: The Merging of Humans and Machines into a Global Superorganism," 1999.

[3] By Ferdowsi, 1010AD. See for example, *The Epic of Kings: Shahnameh*, ISBN 9798637797431, 2020.

[4] By Attar of Nishapur, 1177AD, is a philosophical poem in the teachings of Sufism. See for example, *The Conference of the Birds* (Penguin Classics 1984. ISBN 9780140444346).

Figure I-1. Simorgh as depicted on this book's cover.

The emerging modern Simorgh is a consequence of human civilization and technology. We can attribute traits to it such as beauty, wisdom, and power, as in "The Epic of Kings," and it will include a select population of humans as in "The Conference of the Birds."[5]

Simorgh is a direct extension of human ingenuity, and currently, we are observing the early stages of its development. Simorgh may live, think, and act on new scales of time and space that will be incomprehensible to humans. Many of us will only be functioning as its body cells (nodes) and may be happy with the comfort and security that it will provide (Figure I-3). This splendid development is not likely to stop in the absence of a catastrophic event, and perhaps it is not wise to try to impede its progression. We are at an inflection point

---

[5] The evolution of Simorgh should not be confused with the Gaia Hypothesis of Earth as a living organism that is claimed to have been self-regulating the environmental conditions suitable for life on the planet. The Gaia Theory states that the organic and inorganic components of planet Earth have evolved together as a single living, self-regulating system. It suggests that this living system has automatically controlled global temperature, atmospheric content, ocean salinity, and other factors to maintain its own habitability. In a phrase, this means that "life maintains conditions suitable for its own survival." In this respect, the living system of Earth can be thought of as analogous to the workings of any individual organism that regulates body temperature, blood salinity, and so on. For instance, even though the luminosity of the sun—the Earth's heat source—has increased by about thirty percent since life began almost four billion years ago, the living system has reacted to maintain temperatures at levels suitable for life.

in this development and it behooves us to engage in some conscious early analysis and planning. We need to understand the possible outcomes and consciously influence the emergence of Simorgh in our favor while we can, as our influence will diminish over time. In addition to optimizing its foundation to ensure friendliness to humans, other aspects of the emerging superorganism are currently worth exploring, such as its multiplicity, replication, diseases, and chances of long-term survival.

(a)

(b)

Figure I-2. (a) The nation of birds in conference to seek a leader,[6] (b) The remaining 30 birds that eventually find Simorgh collectively in themselves.[7]

---

[6] Virginia Collera, "Birds Looking for Answers," The Dispenser, January 1, 1970.

[7] B N Goswamy, "What the Bird Prophesised," The Tribune, November 25, 2018.

This is a revolution that is growing faster and larger in scale than anything that humanity has ever experienced or has ever been able to imagine. As we are continuing to merge with our technology, we are going to be catapulted into a fresh reality that will be entirely novel to every human-being alive today.

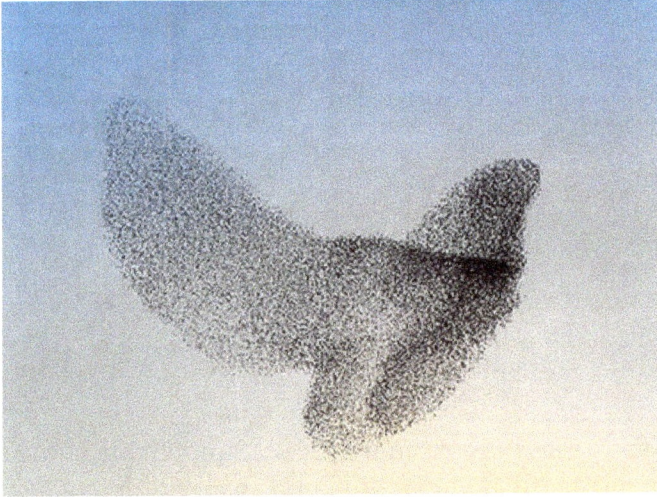

Figure I-3. The outline of this huge bird was formed by a flock of starlings apparently in response to the presence of a predator.[8] The shape was likely formed by chance, but the picture illustrates the concept of a single organism made up of cells that are all individually complex organisms themselves.

Once fully formed, neither the humans that constitute Simorgh's body cells nor those people who live on the outside will necessarily be cognizant of the presence of Simorgh as a living entity. Naturally, the forces and mechanisms that drive its behavior remain equally incomprehensible. This is analogous to the relationship between the human body and the single cells or single-celled organisms that live in and around it. They are not conscious of the human body's existence and intents, even though they are directed or strongly influenced by it (Figure I-4).

Before Simorgh reaches that stage, we currently have a short window

---

[8] Grace Murano, "12 Most Amazing Bird Formations - Murmuration, Amazing Birds," Oddee, September 12, 2012.

of opportunity through which we can exert some influence on the outcome.

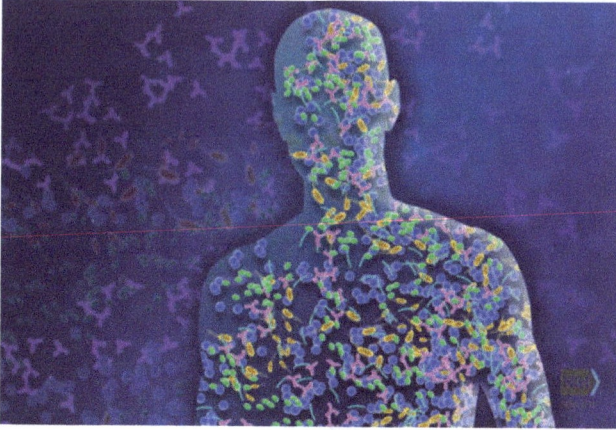

Figure I-4. Cells that live inside and around a human body are not aware of the presence of the human. Source: Darryl Leia, National Human Genome Research Institute.

We will show that with the emergence of Simorgh the human society will simultaneously undergo profound changes. The following are a few examples:

1. The distinction between democracy and dictatorship will fade as both systems become obsolete.

2. The boundary between reality and fantasy will become less distinct.

3. Human nodes of Simorgh will each have the option to design the world in which they prefer to live.

4. Egalitarianism as a general goal will fade away and the human society will permanently bifurcate.

These changes are neither inherently good nor bad. It is up to us to steer them in the direction that appeals to us.

Chapter 1: Stages of Complexity

A commonly accepted definition of a living organism is one that feeds, grows, and reproduces. Other characteristics such as response to stimuli and having a life cycle are sometimes added for additional clarification. Some organisms such as viruses may be considered borderline living since they are not capable of reproducing independently. Cellular life emerged on Earth almost as soon as the early heavy bombardment by meteorites subsided. This was about a billion years after the formation of Earth. Before the emergence of cellular life, there were simple molecules that joined together to create more complex molecules.[9] It has been shown[10] that simple molecules, such as hydrogen cyanide (HCN) and hydrogen sulfide (H_2S), in the presence of ultraviolet light can generate nucleic acid precursors and the starting materials for amino acids and lipids, which are the building blocks of life. Both gases existed in Earth's early atmosphere when ultraviolet light from the sun was abundant. These molecules then combined and formed self-replicating macromolecules such as ribonucleic acid (RNA). Replication established a baseline from which new robust combinations could be created, and natural

---

[9] We don't include molecules that accumulate to form crystals by nucleation because this is a purely mechanical formation that doesn't increase complexity.

[10] Bhavesh H. Patel et al., "Common Origins of RNA, Protein and Lipid Precursors in a Cyanosulfidic Protometabolism," *Nature Chemistry* 7, no. 4 (2015): pp. 301-307.

selection became the mechanism for propagating successful combinations that could replicate well and weeding out defective ones that could not.

Natural selection was the only "programming" that all living organisms received. Inadvertently, it prompted them to strive for the survival of the species. They learned by reinforcement, through a trivial system of reward and penalty: Survival was the reward and demise the penalty. This simple program, in the presence of competition, is responsible for all the traits and individualities that more complex organisms have developed over the millennia. Consciousness, love, self-identity, and problem-solving capabilities are the latest consequences of the same basic programming.

Early cellular life emerged in the form of a cytoplasm enclosed by a membrane. It contained ribosomes and genetic material, but no nucleus and no membrane-bound organelles. Such cells are called prokaryotic cells. They still exist in large numbers today in the form of bacteria and archaea.[11] The start of microbial photosynthesis mainly by oceanic cyanobacteria (prokaryotic blue-green algae) transformed the Earth's atmosphere from being rich in carbon dioxide to having a high oxygen content. This transition is called the great oxygenation event (GOE), which occurred about 2.3 billion years ago (Figure 1-1).

> Consciousness, love, self-identity, and problem-solving capabilities are the latest consequences of the same basic programming.

Initially, oxygen was hostile to all life and could cause damage to living cells. Some cells such as mitochondria learned to harness oxygen for energy production and were able to generate significantly more energy aerobically. Shortly after oxygen's new dominance, and in response to it, some cells started to combine to become larger cells that incorporated mitochondria and other organelles. This

---

[11] Archea are prokaryotic cells similar to bacteria, which are typically found in extreme environments such as hot springs.

more efficient aerobic energy generation was able to support the larger combined body. The drive toward this type of integration is known as biosymbiosis, suggesting that the joined organisms had an improved chance of survival. The resulting "eukaryotic" cells formed membranes to protect themselves from oxygen damage and a nucleus to protect their genetic material, thus the name "eukaryotic" (having a nucleus).

Figure 1-1. Deposits of iron oxide in localized bands in rock formations. They are relics of the emergence of oxygen that caused large scale oxidation of iron in ocean waters. The onset of the great oxygenation event (GOE) is dated to approximately 2.3 to 2.6 billion years ago. Source: Carlos Alberto Rosiere, Federal University of Minas Gerais.

Eukaryotic evolution happened only during a relatively short period of time in Earth's history and did not continue beyond that point. This means that it was only in response to a unique environmental inducement or stress. The great oxygenation event uniquely meets the criteria for such an inducement. GOE not only made aerobic energy generation possible and favorable, but it created the need for the protection of the cell from oxygen damage. Cells with stronger protective membranes and cell walls survived. This extra protection is most likely what restricted additional and alternative eukaryotic cell formations. As a result, all eukaryotic cells on Earth have similar building blocks.

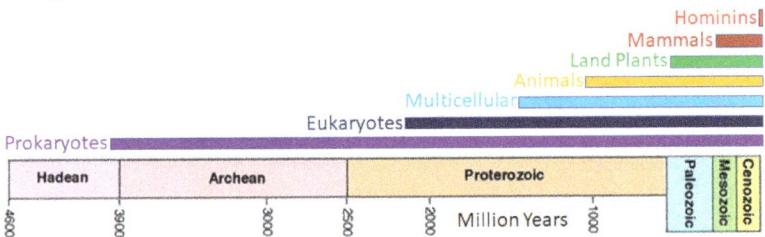

Figure 1-2. Life-related stages of complexity on Earth. Each stage was achieved by the joining of previous elements and the refinement of the outcome.

The next major evolutionary step occurred when various types of individual cells started to combine to form multicellular life (Figure 1-2).

An example of an early-stage multicellular formation can be seen in cells known as choanoflagellates.[12] These are cells that have grouped together to form sponges.

> Individual cells only respond to local stimuli and make no decisions on behalf of the larger organism.

Individual cells must propel themselves in water in order to encounter and capture bacteria; in contrast, groups of cells cooperate to cause the water to move, and capture the bacteria from the moving water. This feeding mechanism is much more efficient (Figure 1-3).

Coordinated group activity is the first step toward the development of a multicellular organism, but it is a reversible process and doesn't necessarily lead to the evolution of a new species. One of the first signs of a permanent transformation is when individual cells respond only to local stimuli and loose the capability to survive in isolation. Other steps beyond this, involve the differentiation and specialization of cells and the formation of system-wide functions such as blood circulation and the nervous system.

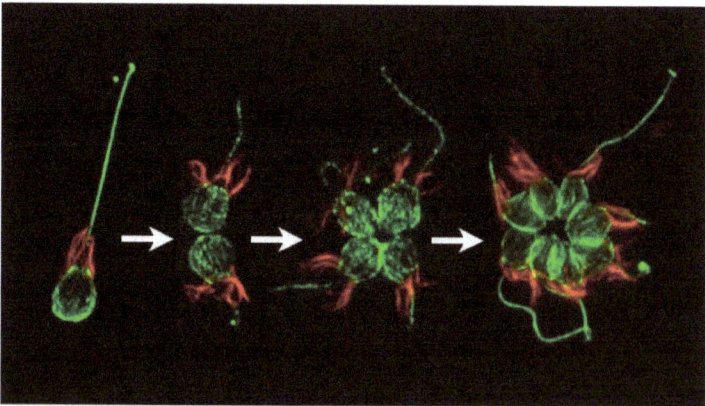

Figure 1-3. A single choanoflagellate cell and multiple cells forming a group. Source: Nicole King, Howard Hughes Medical Institute.

---

[12] "Nicole King," Nicole King | Research UC Berkeley, September 1, 2017.

Communication

Single cells tend to release chemicals in their environment that may be picked up by other cells and trigger group responses such as spore formation in bacteria and yeasts. This communication mechanism has survived and evolved further in multicellular organisms. Communication within a multicellular organism is done either in a systemic (broadcast) mode or in a targeted (point-to-point) mode. Broadcast mode involves the releasing of chemicals such as hormones, to carry each message to all cells of the organism. In contrast, point-to-point communication is done through a chain of neighboring cells mostly by electrochemical stimulation.

The nervous system evolved as a long chain of such neighbor-to-neighbor interactions. Early worms in the ocean were the first organisms to form a basic nervous system. As more nerve cells accumulated on one side of the worm, it started to function as a "brain." Current complex organisms, such as animals, take advantage of both types of communication methods, employing both an endocrine[13] system and a nervous system.

Multicellular organisms gave rise to sponges, fish, and other marine animals, and eventually to land animals that initially crawled out of swamps. One noteworthy swimming organism that predates most current marine animals is the jellyfish (Figure 1-4). The jellyfish has no brain and no blood circulation. It is the ultimate example of how reasonably complex life forms can survive by relying on reflex alone without any brain or central decision making.

> The evolution and success of complex organisms did not cause the demise of simpler life forms.

Another important point to emphasize is that the evolution and success of complex organisms did not cause the demise of simpler life forms. In other words, "survival of the fittest" doesn't have a unique outcome. Early life forms such as bacteria and archaea

---

[13] Discussed in more detail in Chapter 15.

have survived to the present day, and many types of bacteria have symbiotic relationships with the complex organisms that they are the ancient ancestors of.

Figure 1-4. As a primitive form of marine life, the jellyfish has no brain, no heart, and no skeleton. Source: Natursports, Dreamstime.

Latest Stages of Complexity

Some stages of complexity began with a group of individuals gathering synergistically to achieve a goal that was not reachable by each individual. As their cooperation became closer and more permanent, they gave rise to a new organism with more advanced capabilities. Each individual's impetus to join the group was to reduce their burden of self-support and improve their chances of survival.

Measured by their level of intelligence, humans are thought of as having assumed the highest level of complexity and capability among all animals. Furthermore, humans took steps towards yet another stage of higher complexity by forming civilization that achieves goals beyond the capability of each individual. Human civilization has led to the development of technologies that are linking people more closely and permanently, and each person's role in the society is becoming more focused and specialized. This is the recipe for some human societies to begin assuming the identity of a new organism: the identity of Simorgh. Civilization and technological advancements have continued to make people's lives easier, and their chances of survival have steadily improved. The continuation of this trend is the motivation that will ultimately enable the formation of Simorgh. As in previous emergences of eukaryotic cells and multicellular organisms, this development may not encompass the entire society. Many people are going to opt out of or be excluded from Simorgh incorporation, even though the presence of Simorgh can be beneficial to all.

Summary

Simpler entities tend to merge into more complex units under the right circumstances. Simple molecules joined to make complex molecules; complex molecules joined to make self-replicating macromolecules that then joined with others to make prokaryotic cells. Then, in response to changes in the atmosphere, prokaryotic cells joined to make eukaryotic cells. Multicellular organisms and complex animals followed. In each stage, it was always a small subgroup that ascends to the next level. The rest retain their level of simplicity indefinitely. Humans took the first steps toward the next stage of complexity by gathering in cities to form the human civilization. Technology is helping to bring this trend to fruition by the formation of Simorgh within subgroups of the human population.

Chapter 2: Civilization, Culture, and Government

Humans have always been semi-social animals: Unlike solitary cats, we need to form long lasting family and tribal bonds, and unlike bees, we value our individual freedoms, and we don't always tend to do what is best for the society. Early humans naturally tended to live in small families and bands where interpersonal ties could coexist with selfish behavior. Disputes and quarrels among individuals would therefore be common but manageable. Such disputes were often settled by force, as can still be seen in any troop of primates today. A strong figure such as an alpha male could forcefully maintain order among the group. As human communities grew larger, certain lasting principles and norms were established to reconcile individualistic aspirations with community preferences and to prevent destructive outbreaks of violence.

Early tribes relied heavily on strong family ties to function, and decisions were frequently made through multilateral negotiation. A government of sorts appeared when the size of the tribe grew larger and community-wide decision making was needed. Wiser and elderly tribe members would be selected to adjudicate disputes and serve as general advisors to the tribe. When tribes grew into chiefdoms, and permanent settlements started to take shape, the additional tasks of infrastructure building and food distribution required central planning. Multiple individuals or councils were needed to serve in planning, judicial, and advisory roles to accommodate the needs of the larger population. Eventually, the councils grew larger, and became the city government. Initially, there was a certain degree of flexibility in this type of government, and the collective will of the population played a major role in policy adoption. An elected central figure, or ruler, often coordinated government functions by consulting with the populace. This form of "direct vote democracies" tended to be short lived. Over time, the governing body would become an elite group within the society and would learn to maintain its grip on the reins over successive generations.[14]

Another type of government formation that eventually became more common was for a warrior or a military leader to take full control of the society and assume ownership over it. The ownership concept meant that the ruler could give his "ownership" of the territory he ruled over to his designated heir. This concept formed the first ruling dynasties. This type of government is referred to as an autocracy, monarchy, or a dictatorship, suggesting that one single person had unlimited power over the society. In reality, the one-man rule was only in appearance because the job of ruling over a large population is beyond one person's abilities. Typically, an oligarchy would form, consisting of people who were loyal to the ruler or indispensable for managing government functions. These people formed the "ruling class" which was typically a closed class in the society, meaning that ordinary people would find it difficult to enter this class. Transition of power from one generation to the next could be peaceful if one uncontested successor was identified; otherwise, violence erupted in the absence of a consensus. Succession by popular vote that some societies adopted[15] was primarily intended to ensure a peaceful transition of power.

---

[14] Morton H. Fried, The Evolution of Political Society: an Essay in Political Anthropology (New York: Random House, 1967).

[15] For example, in the post-classical period in the Indian subcontinent, it was common for rulers to be elected by the vote of influential people. One case was the election of Gopala, the founder of the Pala dynasty.

The rule of law was a concept promoted by some early rulers who sought to provide a common basis for the judicial system and for conflict resolution in the country. This was to replace subjective judgements by "wisemen". One of the earliest written sets of such laws is credited to the Babylonian king Hammurabi of the eighteenth century BC.[16] Other such historical laws include the Ten Commandments by Moses and possibly the first declaration of human rights by king Cyrus of Persia (Figure 2-1). These three sets of laws are between twenty-five to thirty-eight centuries old.

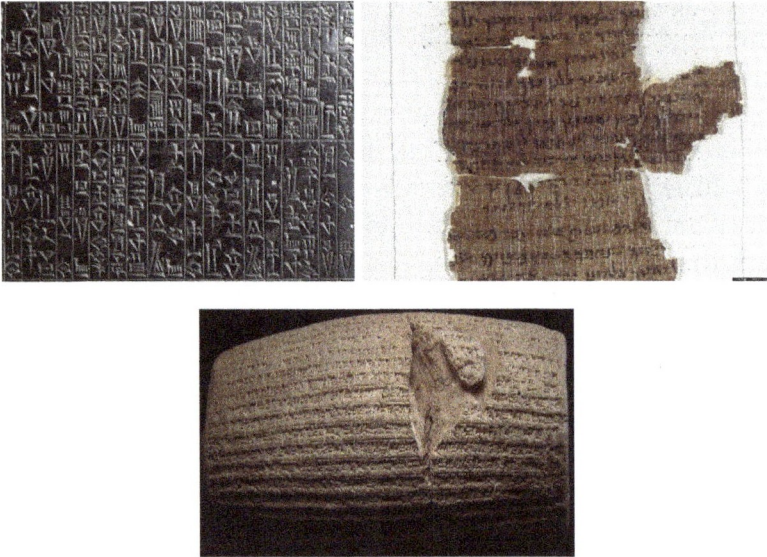

Figure 2-1. From left to right: The code of Hammurabi in cuneiforms,[17] an old manuscript of the 10 Commandments,[18] and the cylinder of Cyrus believed to contain the first written proclamation of human rights.[19]

---

[16] Hammurabi code of laws, written on a large slab of black stone that's currently at the Louvre Museum. It consisted of 282 rules of social and commercial transactions.

[17] History.com "Code of Hammurabi," History.com (A&E Television Networks, November 9, 2009).

[18] By Megan Gannon, "Ancient Copy of Ten Commandments Goes Digital," NBCNews.com (NBCUniversal News Group, December 13, 2012).

[19] Currently in the British Museum, it proclaims freedom of religion among other provisions.

As mentioned above, "direct-vote democracies" were rare, and limited to small city-governments. In direct-vote democracies, people would have to gather to vote on most major decisions. As the population grew, this approach became less feasible. Still, many larger societies could be governed in a similar fashion, as "representative democracies," where decisions were made by people's representatives rather than by direct popular vote. The word "democracy" suggests government by the vote of common people; however, the intention of the promoters of democracy was never ruling by ordinary people or "mob rule," as this would be destabilizing to the society. In a democracy, minority rights need to be protected by strong laws and civil institutions. Government responsibilities require the involvement of highly trained and specialized personnel, and the general population has neither the necessary skills nor the sound judgement to decide the policies and much of the day-to-day business of the State. The general public is at best qualified to vote on local issues that directly affect their daily lives, such as building a local stadium or local roads, or to elect trusted local representatives to make decisions related to the broader government responsibilities.[20]

> The general population is at best qualified to vote on local issues that directly affect their day to day lives.

The presidential election process in the United States was not originally designed to be carried out directly through popular vote. Instead, each state would send "electors" to a temporary congregation known as the "electoral college" which was tasked with electing the president and the vice president for the country. Alexander Hamilton, a U.S. founding father, argued in support of the electoral college system by stating that electors could have access to essential information unavailable to the general public.[21] The media, on the

---

[20] This is partially related to the fact that people are generally more interested in narratives than in the truth, unless it pertains to their immediate surroundings and daily life. See Chapter 6.

[21] Alexander Hamilton, "The Federalist Papers: No. 68," The Avalon Project: Federalist No 68.

other hand, prefers the fanfare and excitement of elections by direct popular vote between opposing candidates. Direct elections take the form of a popular sports match, and some people vote for the political party that they adopt as their "home team" instead of voting on the merits of the candidates. In retrospect, an electoral college system without advertised presidential candidates would better serve the nation by electing more qualified statesmen as leaders. People would vote for local electors based on their qualifications and local reputations, rather than presidential candidates whose true characters may not be so transparent. The electors in turn would have enough time to study and understand the qualifications and goals of each presidential candidates before voting for the candidate of their choice. The electors would be less susceptible to demagoguery and misinformation than the general public.

Despite its flaws, representative democracy allows social mobility into and out of the otherwise closed ruling class. An ideal society is not a classless one, but one that has fully open classes[22] with significant mobility among them. Representative democracy opens one class, but it is not government by the people, as it should not be. Mob rule is prevented by filters such as political parties that screen the qualifications of candidates before elections. Checks and balances are built into the system by dividing the government into multiple branches. The intention is to prevent corruption and favoritism. However, these systems still have inherent instabilities that tend to move them toward minority rule. Political parties themselves are prone to corruption, and extremist demagogues can induce fear in the population in order to enhance their chances of election and maintain their

It can be argued that a further split in government functions between domestic affairs and foreign affairs of the executive branch would be beneficial.

---

[22] An open social class is one that people can enter or leave by following accessible procedures and acquiring proper training. Other than the ruling class, two other classes that remain closed in many societies are the extremely rich and the extremely poor classes. One has no standard procedure for entering, and the other has no standard procedure for exiting.

hold on power. Some of these flaws, as discussed in the following sections, can be mitigated by increased automation of government functions and the implementation of instant voting.

Typically, a modern government is split into three branches: legislative, judiciary, and executive. This tripartite was first suggested by Baron de Montesquieu (Figure 2-2) in his classic book *The Spirit of the Laws*. He suggested that the best way to prevent despotism and corruption was through the separation of government powers into the three listed above, with all these coequal bodies subject to the rule of law.[23] With recent internationalization of lifestyles, it can be argued that a further split in government functions between domestic affairs and foreign affairs of the executive branch would be beneficial. People in charge of the independent foreign affairs branch could be appointed by the legislature. This arrangement would shield a country's sensitive foreign policy from the volatile domestic politics and would be more conducive to international cooperation.

Figure 2-2. Great Educator: Baron de Montesquieu 1689 to 1755 to whom the concept of the separation of powers in modern governments is attributed.[24]

---

[23] Montesquieu Charles de Secondat, The Spirit of Laws Translated from the French of M. De Secondat, Baron De Montesquieu. A New Translation. In Three Volumes. (Berwick: Printed for R. Taylor, 1770).

[24] "Great Educator: Baron De Montesquieu 1689 To 1755," Ragged University, November 24, 2016.

We are currently in the midst of a gradual but far-reaching shift in the way government operates. This is brought about by the penetration of smart computer algorithms into government functions, legislation, and decision making. Algorithms can take into account all relevant background details, policy tradeoffs, and even the weight of public opinion. No one with a limited time will be able to challenge the well-planned automated policy recommendations, and soon most government decisions will become algorithm-driven. This new paradigm is likely to be more efficient and less susceptible to corruption assuming that the algorithms are properly scrutinized and any built-in biases are removed over time.

> Most government decisions will become algorithm-driven. This new paradigm is likely to be more efficient and less susceptible to corruption.

Government by Algorithm

In the early twenty-first century, the European country Estonia became a pioneer in building automation into government functions and replacing government bureaucracy with algorithms and AI.[25] All citizens were assigned secure digital identities that allowed them to conduct much of their governmental and private business online. Blockchain technology safeguarded personal and public information and made them tamperproof. Such protected information included health records, legal documents, court system records, police data, banking data, business information, and real-estate registries. All business including voting is completed with a few keystrokes without the need to reenter personal information.

---

[25] "Estonia - We Have Built a Digital Society and We Can Show You How," e-estonia, February 25, 2021.

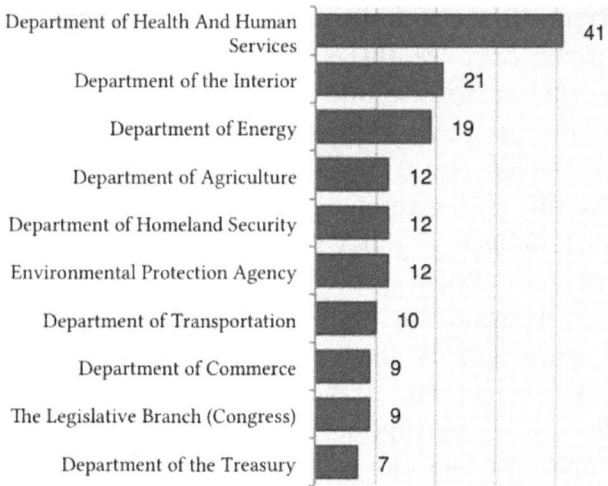

Figure 2-3. An informal assessment of the relative reliance of various US government agencies on algorithms in 2016. Source: D. Trielli, J. Stark, N. Diakopoulos, "Algorithm Tips: A Resource for Algorithmic Accountability in Government," Northwestern University.

Similar trends have developed, albeit with a lag, in other industrialized countries. An increasing number of government functions are performed by computer algorithms that process tax documents, applications, and surveys (Figure 2-3).[26] Applicants can complete forms online where many fields are auto-filled based on the applicant's identity. The applications can be automatically processed without much human involvement. In essence, the applicant simply introduces himself to the system and requests a service. The system quickly determines eligibility and responds to the applicant.

Using artificial intelligence, and "natural language processing," the transition from keyboards to voice-based virtual client-support is accelerating. The client will converse with a virtual agent who may assist with financial transactions, or issue documentation such as a passport, a travel visa, or an export license (Figure 2-4). Contractual and ownership documents that are currently enforced by the court

[26] David Freeman Engstrom et al., "Government by Algorithm: Artificial Intelligence in Federal Administrative Agencies," Artificial Intelligence in Federal Agencies, February 2020.

system will be replaced by secure blockchain ledgers, or "smart contracts." With smart contracts, conditional financial transactions may be executed automatically upon fulfillment of the conditions. Thus, escrow accounts, for example, would not be needed in most cases.

Eventually, completing forms and providing any personal information will become outmoded everywhere. The client will be automatically identified by the system, and will only need to request services, such as permits, loans, judgements, etc. and the request will be processed based on the applicant's known identity and social records. There is no reason for processing to not be almost instantaneous. Overtime, kinks and flaws that inevitably exist in all early automations will be identified and removed, and quality of service will steadily improve.

Extrapolating this development, we see that in the first step, instead of applying for a service, people simply ask for it. In the next step, even asking will not be necessary. Needs are automatically identified and processed by digital assistants.

> By the time bureaucracies are reduced to such levels, humans who participate in the system will have become body nodes of Simorgh.

By the time bureaucracies have been reduced to such levels, humans who participate in the system are on their way to becoming body nodes of Simorgh. Their reasonable daily needs and desires will be met automatically with little or no effort on their part.

Budget allocation and approval are other major government functions that are carried out very inefficiently today. In most countries, the process is infested with favoritism, influence peddling, and corruption. Automation can start to improve the process. We know from examples such as the government of Estonia that once a general budget has been approved for a region, the implementation of the budgeted tasks can be managed much more efficiently by algorithms than by human bureaucracies. Annual adjustments to the budget are needed based on changing demographics and business activities.

Such adjustments can also be carried out precisely and efficiently by algorithms and AI, with the help of automated feedback from the population. In this way, algorithms can start with a one-year human-processed budget and work out budget changes for subsequent years without the need for human decision makers.

Government by algorithm takes a great burden off the shoulders of government employees and elected officials. Algorithms can stay impartial, conduct the tasks without any mistakes, be available at all times, and not suffer from boredom or fatigue, all of which will become the norm in the society as dependence on human involvement declines.

Figure 2 4. The character Cora was introduced by Natwest Bank as a digital-human customer service representative. Such audiovisual "Chatbots" will be able to facilitate the delivery of many government and private sector services to customers. Source: "Ask Cora," natwest.com.

In addition to routine procedural tasks, more complex decision-making algorithms such as "judgement automation" are currently being experimentally deployed by developers around the world. They range from determining eligibility for government services, as discussed above, to the deployment of special law enforcement personnel to various locations throughout a city, as well as the resolution of minor disputes.

One may think that the management of social or political crises are beyond the reach of automation, but it is not. Social models have already been integrated into algorithms that can predict the emergence of a crisis based on early signs and can help manage the crisis by presenting options to the authorities.

Listed below in blue are the major government functions that have the immediate potential to be performed by algorithms:

- Updating Laws and Maintaining Order
  - Property Rights
  - Justice
  - Policing
- Providing National Defense
  - Foreign Policy Formulation
  - Maintaining Armed Forces
- Infrastructure Building
  - Management of Large Projects
  - Transportation Improvement and Planning
  - Protection of the Environment
- Promotion of Economic Stability and Prosperity
  - Facilitating Commerce
  - Investment in Research and Development
- Revenue Generation
  - Taxation

Table 2-1. A sample list of some government functions. All items have the potential of being automated. Automation is already penetrating items highlighted in blue, while purple item will take somewhat longer to automate.

One of the major advantages of using algorithms in lawmaking and in large government projects is the possibility of simulating the project's impact on society before implementation. For example, the effects of a minimum-wage increase may be simulated first to see if it may have unintended consequences such as eliminating certain jobs. Or the effect of building a stadium in a neighborhood, for which various effects such as traffic bottlenecks can be simulated and observed ahead of time. Simulation may not be able to predict the exact outcomes of projects and legislation, but it can draw attention to potential side effects prior to enactment. Efficacy can be ensured, and unintended negative consequences may be minimized.

It is a matter of time before the entire government will be algorithm-driven. Human heads of state, if any, will only be figureheads. Some people express concern about various social biases that may be built into algorithms, either inadvertently or deliberately. This is undoubtedly true in some cases. However, it is worth noting that biases in algorithms are much more easily discovered and corrected over time than prejudices and irrationalities held by human administrators.

> Decision making without human guidance signals the dawn of the age of Simorgh: the age of humans merging with their technology to create larger and more complex entities.

There is always an algorithm or a method behind any type of decision making (Chapter 12). When you explain how you chose among multiple options, you describe the personal method you used in your mind. Even emotions or mental biases may be interpreted as quick-response algorithms. AI uses a set of algorithms that can learn how to achieve certain goals. The methods it finds may not be known to the programmers in advance, but the goals are always known and clearly specified.

The difference between decision making by an individual vs AI is the difference between private and public algorithms. AI offers more transparency in that its goals and algorithms may be examined and improved over time, while mental algorithms used by individuals are not transparent nor subject to scrutiny.

Decision making without human guidance signals the dawn of the age of Simorgh: the age of humans merging with their technology to create larger and more complex entities. We are currently observing the emergence of an early Simorgh who is learning to make decisions on its own without human supervision.

Automation, AI, and Democracy

A benevolent dictatorship, or the rule by a "philosopher king," to use Plato's nomenclature, has long been considered to be one of the most effective forms of government. Major decisions

> The distinction between a benevolent dictatorship and a democracy is fading.

are made quickly and efficiently, and they are good decisions because of the well-intentioned and well-informed ruler. There have been many autocratic rulers in history who were considered benevolent by the majority of their subjects.[27] Obviously, they exercised good judgement and implemented policies that benefited most people. However, there are two major problems with advocating this form of government: One is that it is hard to find such high integrity rulers, and even if found, they are likely to be corrupted by power over time. The second problem is one of succession. After the loss of the leader, the benevolence of the dictatorship would not necessarily persist. Therefore, a benevolent dictatorship would only be ideal if the ruler 1) were to stay responsive to the people's needs and wishes at all times; 2) were immune from corruption by power; and 3) would never pass away. All this may sound whimsical, but it is intriguing to note that a government by automation and AI would potentially satisfy all the ideality requirements listed above. Taking the first condition of responsiveness to people literally, would imply that AI-assisted government could even have elements of a direct-vote democracy. Responsiveness to people is a prized hallmark of democracy. With the help of technology, the distinction between a

---

[27] Marcus Aurelius of Rome, Khosro I of Persia, Yang Jian (Wen) of China, Lee Kuan Yew of Singapore, and George Washington of USA are a few of the rulers generally referred to as benevolent dictators.

benevolent dictatorship and a democracy is fading.

Many provisions of the constitutions written for representative government were designed for the means and methods of public engagement available a few centuries ago. Most election and voting laws have not been sufficiently updated to take advantage of current technology. For example, in the United States, people vote infrequently and only on announced dates, while today, it is possible to know people's preferences in real time without the need for a national voting day. Binary yes or no votes on many issues are not representative of the preferences of the population. Voting on a zero to ten scale, for example, would convey more information, but it was too difficult to count in the past. The concept of "one person one vote" is not ideal, as it does not differentiate between the votes of those who care about the subject and those who don't. The vote of a person who has studied and deliberated a subject has the same weight as that of a person who is simply responding to fake news and propaganda. Trying to attach weights to people's votes has not been practical and would naturally lead to inaccuracies. An impartial electronic system on the other hand can add an appropriate weight to any vote, based on the amount of knowledge and passion behind it. Such attributes can be judged with surprisingly high accuracy by asking the voter to respond to a few engaging questions. Even in the absence of any such questions, there is already ample information on people's public profiles to show their areas of interest (or lack thereof). Ultimately, real-time opinion gauges, based on every citizen's daily discourse, can provide a very reliable snapshot of popular sentiment on any social issue. No voting will be necessary.

An AI-based modern democracy can:

1. Eliminate the need for voting because everyone's social preferences are always detectable from their online activities and the questions they may be willing to answer.

2. Replace a representative democracy by a direct vote democracy even in a very large country, because sensing people's sentiments has become rapid and effective.

3. Blur the distinction between a democracy and a lasting benevolent dictatorship. The AI-based government will be very responsive to the population's social preferences.

Summary

Government and the rule of law provide the mechanisms for the cohesiveness of the society. Increasingly various government functions are becoming automated and algorithm-driven. Automated government functions are much faster, more efficient, and less prone to errors. By the time the society transitions into Simorgh, government decisions will no longer be made by individual humans. The rule by AI will have the best characteristics of a benevolent dictatorship and a direct-vote democracy.

Chapter 3: International Disputes and War

The part-social, part-individual nature of humans has always led to conflicts between selfish aims and collective interests within the society. Prior to civilization, early humans resolved many such disputes aggressively like most primates.[28] Those who were physically stronger would of course win the argument. Civilization created other more peaceful means of conflict resolution such as arbitration and the courts of justice. This did not eradicate violence in the society, but reduced its level considerably. City dwellers to this day typically do not resort to storming their neighbors' houses to resolve disagreements. Such disputes are settled in the courts, according to accepted legal guidelines. Unfortunately, the creation of civilization also gave rise to a more ferocious and destructive form of violence, albeit less frequent, between city states and countries. This came to be known as war, and it is a menace that human societies have never been able to quell.

> The creation of civilization gave rise to a more ferocious and destructive form of violence, between city states and countries.

---

[28] Evidence of violent death has been found in ancient human skulls and skeletons dating to as early as 400,000 years ago. Deborah Netburn, "430,000-Year-Old Skull Suggests Murder is an Ancient Human Behavior," Los Angeles Times, May 29, 2015.

Most of us are not fully aware of the horror and savagery of war. Unfortunately, in many narratives and history books, wars have been considered justified or even romanticized as clashes between good and evil, and their barbaric nature has been underreported. From the early days of civilization, wars have been extremely violent and destructive. Dwellings and city buildings were often set on fire or otherwise totally obliterated. Nevertheless, the domain and spread of ancient wars remained limited thanks to people's restricted mobility at the time. The destructive power of war grew rapidly after the introduction of explosives to warfare in the fifteenth century, marked by the battle for Constantinople in year 1453. In the twentieth century, World War I claimed a direct and indirect death toll of over forty million and became one of the deadliest wars in human history. The alarming level of death and destruction prompted many to declare World War I as "the war to end all wars" and to devise alternative means of international conflict resolution that excluded war. In pursuit of this goal, an international body called the League of Nations was founded. It declared that an attack on any member nation would be considered an attack on all. This promise, together with the newly formed International Court that could be the arbiter of justice in any cross-border dispute, was to obsolete war as a means of conflict resolution between any two countries.

The League of Nations created a new sense of internationalism and performed useful international functions, such as promoting trade and cooperation, but it proved to be ineffective in preventing major wars. Its lofty motto of a united response to aggression proved to be impractical, as no nation stepped forward to prevent the hostile international campaigns that ultimately led to World War II just two decades later. This war was even more destructive than the First World War, with casualties exceeding fifty million. The sentiment to "end all wars" grew stronger,[29] and by the end of the Second World

---

[29] This statement about war is attributed to Albert Einstein: "May the conscience and the common sense of the peoples be awakened, so that we may reach a new stage in the life of nations, where people will look back on war as an incomprehensible aberration of their forefathers." Albert Einstein and Alan Harris, The World as I See It (New York: Philosophical Library, 1949).

War, a presumably more "practical" United Nations was founded to replace the League of Nations (Figure 3-1). The United Nations also has an International Court of Justice for conflict resolution and many other branches to encourage economic and humanitarian cooperation among nations.

Despite its survival, the United Nations cannot take much credit for the fact that there has not been a major multinational war since its inception. The single factor that has prevented such major clashes is the concept of "mutually assured destruction" in the aftermath of the use of nuclear weapons in World War II and the ensuing Cold War. Smaller bilateral wars and proxy wars have continued to devastate many regions of the world, and this primitive and savage behavior prevails. One main reason for its persistence is that major military powers of the world show no interest in empowering such organizations as the International Court of Justice, and therefore international law and order cannot take a foothold. Another enemy of peace is the lucrative international arms trade, which currently exceeds 400 billion dollars per year,[30] as well as the massive industrial complex that ever-increasing military expenditures support.

One step that may be taken to pave the way for an eventual elimination of the specter of war is for every country to adopt two distinct executive branches of government: one for domestic affairs and one for international affairs, as discussed in Chapter 2. Once the role of the volatile domestic politics is minimized in international relations, collaboration agreements and treaties among nations to enforce international law will have a higher chance of success.

Cooperation Blocs

The concept of world unification is attractive and is different from having a single world government. Unification is a special case of forming cooperation blocs among nations as we see today. This

---

[30] The top 100 arms companies made an estimated $398.2 billion worth of sales in 2017. "Killer Facts 2019: The Scale of the Global Arms Trade," Amnesty International, August 23, 2019.

sort of cooperation or common market could, in principle, include all nations. Even though it is unlikely to happen, it is a tantalizing possibility. On the other hand, multiple cooperation blocs are very likely to shape the future of international affairs. This has already partially formed, and competing information-sharing blocs of nations have emerged. The formation of such blocs and their multiplicity will have a profound effect on the world's next major evolutionary advance, namely the development of Simorgh.

Figure 3-1. The United Nations headquarters building in New York City on Nov. 17, 2017. United Nations has an annual budget of $5.4 billion and 44,000 staff members worldwide (2018). Source; Brendan McDermid, Reuters.

Advanced War Technology

It may be counterintuitive, but with new technological advances, international conflicts will lose their mass-destructive character. Over the years, advances in explosives technology and the sophistication of delivery systems steadily raised the destructive power of war to its culmination, with nuclear weapons that debuted in World War II. Since then, such nuclear weapons have been deployed on numerous long-range missiles for potential delivery to military targets. Many people feared that the widespread use of nuclear weapons could potentially wipe all human life from the face of the Earth; though the ensuing panic was short-lived as the deterrent capacity of nuclear weapons became apparent. The proliferation of nuclear technology is continuing slowly but steadily, and eventually every nation will have access to it despite severe opposition by those who possess the technology now. Some of the strongest opponents of proliferation are military armament exporters who prefer to retain the high demand for conventional weapons. A country that finds security in its nuclear armament is less inclined to allocate a large budget toward importing expensive conventional weapons. Nuclear proliferation may increase the likelihood of accidental use, but it is more likely to prevent a full-fledged war. It will change the nature of warfare toward less destructive force with higher accuracy. Recent advances in automation and information technology are making precision and targeted strikes more feasible. Consequently, the use of indiscriminate destructive force may no longer be contemplated. This is a continuation of how the cold war was fought between the US and USSR. The concept of mutually assured destruction by nuclear weapons forced both sides to refrain from an all-out clash and resort to covert and targeted operations instead.

> With new technological advances, international conflicts will lose their mass-destructive character.

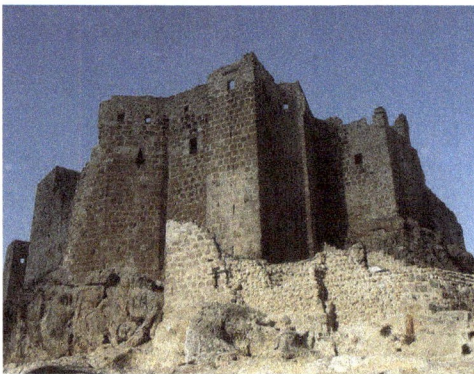

Figure 3-2. The remains of Masyaf castle in today's Syria. It was a stronghold of the religious sect known as the "assassins." Source: sipazigaltumu, flickr.com.

Covert Operations

Over the ages, covert operations - including sabotage, assassination, and private threats - have been an integral part of most international conflicts. Noteworthy in history is the campaign by Saladin, the ruler of Egypt against a powerful extremist religious sect known as the Assassins in 1176 A.D. Seeking their destruction and surrender, Saladin placed a siege on the castle of Masyaf, the Assassins' stronghold, that is located in today's Syria (Figure 3-2). It has been reported that one night as Saladin was asleep beneath the castle, an Assassin managed to infiltrate his tent and left a dagger with some poisoned pastry and a written warning note next to Saladin's bed. Following this episode, Saladin ordered the withdrawal of the siege and sought to make peace with the Assassins instead. This is an example of how covert special operations can prevent or change the course of a war, and minimize the use of massive force.

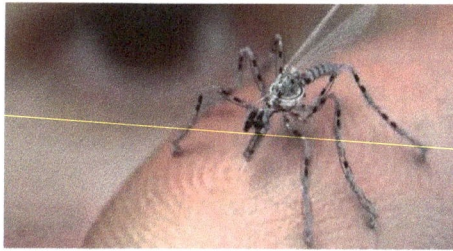

Figure 3-3. Insect Spy Drone is an example of a military MAV. It has a camera and a syringe.[31]

Technology advancement has created new methods of covert action. During the Cold War, the newly developed small cameras, listening devices, and transmit/receive modules were used as aides to human spies and special operation personnel. More recently, many such devices can be placed on autonomous vehicles such as an unmanned aerial vehicle (UAV) or a micro air vehicle (MAV) to carry them to the target. Figure 3-3 shows the concept of a future MAV in the form of a flying insect that employs a camera and a syringe. Such vehicles may be remotely controlled by advanced AI algorithms to collect

---

[31] David Mikkelson, "Insect Spy Drone" Snopes.com, November 17, 2020.

information or to act as a physical threat to the enemy as a part of covert military operations.

Autonomous Action

The general trend in military technology has recently been to move away from human controlled equipment, or remotely controlled gear, to autonomous action. Autonomous weapons technology is advancing at a very fast pace, offering significant advantages in both capabilities and cost to military operations.

Military fighter aircraft, bombers, and spy planes have become more sophisticated and expensive by the day. Extreme speeds and accelerations achieved by fighter planes strain the physical tolerances of human pilots to the limit. UAVs, on the other hand, don't need to support the human pilot, and can engage in maneuvers too taxing to the human body. They can stay aloft for days, and they don't suffer from human fatigue. If there are human operators to remotely control such vehicles, they can switch shifts or divide the responsibilities in order to avoid exhaustion. On the other hand, in tactical engagement situations, the human operator of a remotely controlled UAV may not react fast enough to the threats. Thus, the vehicle would benefit from autonomous decision making, or a nonhuman remote operator. These capabilities are evolving rapidly.[32]

Figure 3-4. Napoleon Bonaparte, Hannibal Barca, and Genghis Khan are known as three of the most capable battlefield commanders of all times. Sources: U.S. Library of Congress. Mommsen's "Römische Geschichte" page 265, *Hannibal*. britannica.com.

---

[32] Northrop Grumman's X-47B is an unmanned combat aerial vehicle (UCAV) employed by the US Navy.

Battlefield management and battle command have the responsibility to plan, direct, and lead forces against an enemy in a hostile environment. In the past, such functions were performed by military commanders and their staff (Figure 3-4). They planned where and how to attack, retreat, and redeploy if needed. Troops and types of weapons were limited, and an experienced commander could visualize the entire battlefield in his mind. In modern times, the physical battlefield is not always geographically contiguous. There are many more weapon types, and events unfold at a much faster pace. The pattern of force deployment and the mix of weapons need to quickly adapt to the maneuvers of the adversary. Today's fast changing conflict theatre can cognitively overload any battlefield management team with the sheer number and frequency of threats and the need for quick responses (Figure 3-5). It is imperative for commanders to rely on electronic battlefield management systems (BMS), which are automated mechanisms of information acquisition and processing to help with command, control, communications, and intelligence (C^3I) functions of a military unit. BMS systems are currently used as aids to commanders, but as they become augmented with AI technologies, many decisions will be taken automatically, to the degree that eventually no tactical decisions by human commanders will be necessary or possible.

> No tactical decisions by human commanders will be necessary or possible.

- Battle Management System (BMS)
- Naval Combat Management system
- Reconnaissance / Surveillance Systems
- Tactical Area Communication System
- Air Defense Early Warning C4I System
- Fire Support Command Control and Communication System

Space
Cyberspace
Air
Land
Sea

Figure 3-5. Today's battlefield is complex and agile beyond the capability of any unaided commander to direct. Some the tools used in battlefield management are listed.

Warlords and Mercenary Armies

With prolonged regional wars in Europe in the Middle Ages, mercenary armies for hire were formed in addition to state-run armies. Any ruler could hire a mercenary army to wage war against an opponent. An infamous example was the White Company mercenary army of the fourteenth century Italy. At the time, the Pope and many fragmented city-states in Italy were at a state of war with each other. The White Company mercenaries took full advantage of the situation by fighting on behalf of the highest bidder in any conflict. They fought both for and against the Pope, the city of Milan, and the city of Florence. The mercenary army was autonomous, so the ruler who employed such an army would leave all the planning and conduct of the battles to the Company and would not manage the details.

In addition to mercenary armies, there are numerous accounts in history of autonomous warlords who ruled over a territory within a country and lived independently of the central government, relying on the military power they wielded in the region. The "warlord era" in China for example was a time in early twentieth century when the country was being controlled by several local warlords who were army commanders of each region.

In the future, the AI-driven military force of any country may act like a mercenary army, in that the government that initiates a war may not have any control over the way it is conducted. It may order a limited engagement, but assuming that the goal is to win, the government can quickly lose the ability to control the scope. Tactical and strategic decisions will be made too rapidly for the government to be able to intervene.

> We encounter a slippery slope when we try to define morality for the AI system that is in control of a military force.

An AI-based military force also has the potential to act as an independent warlord and seek its own self-preservation. There are efforts to incorporate morality and ethics in AI programming such that the sanctity of human life is preserved. However, we encounter a slippery slope when we try to define morality for the AI system that is in control of a military force. Terminating human life needs to be permitted as long as the target is tagged as a foe. Friends and foes may be switched with shifting alliances, so the tagging process needs to remain relatively fluid, while winning and surviving are the main goals. Therefore, if a military force gains enough autonomy to become a fully developed Simorgh, it may be able to tag any human as a foe and become the enemy of humanity. Humans may be essential for Simorgh's survival in the beginning, but their importance can diminish over time.

For us humans at this stage, it is important to make sure that military Simorghs cannot form. There is no international law that can enforce this. Therefore, the only logical way to prevent the formation of a military Simorgh is to eliminate the specter of war from human society ahead of time. This is difficult but achievable by establishing and enforcing a universal rule of law, such as what was envisioned by the founders of the League of Nations and the United Nations.

Summary

The act of war is a lingering remnant of our savage past. Vast resources are being allocated to military technologies aiming at full automation of weapon systems and battlefield management. The first Simorghs that appear on Earth may be autonomous military forces, which may be to the detriment of the Human Race. Morality cannot be robustly programmed into the algorithms that run military operations since morality in war is an incongruity. Therefore, a military Simorgh may become a major threat to humanity. It is in our best interests to eliminate war as a means of conflict resolution as soon as possible, before military Simorghs appear on the horizon and threaten our survival.

Chapter 4: Intelligence Centers and the Control of Human Society

In the context of governmental affairs, the word "intelligence" is a general term used for information gathering on potential adversaries. Espionage is a special case where the information is not public, and could be gathered privately or in an illicit way. Intelligence agents, spies, and information bureaus have existed since the early days of human civilization in order to give the governing bodies knowledge about the population and any potential internal or external threats. Information gatherers who served in a king's court would have to dramatically increase their activities during wartime. Gaining any strategic knowledge about the enemy's capabilities and plans was both difficult and essential.

One of the early texts available on this subject is the Chinese book *The Art of War* written about the fifth century BC by Sun Tzu. Under the topic "Employing Spies," Tzu stresses the importance of "foreknowledge" and he equates it to power. He promotes intelligence as the means to prevent "commotion at home and abroad" and suggests compartmentalization of intelligence gathering in order to maintain central control.

Gathering secret information has its many challenges and dangers, but the transmission or delivery of the secretly gathered information is generally more prone to detection and interception. In the antiquity, all information was, by and large, obtained and carried by human informants. Often multiple humans were involved in carrying written or verbal information from its source to its intended destination. Some non-human help was also used from time to time where human travel was too conspicuous or inefficient. Such non-human information carriers included homing pigeons, floating messages on rivers, and visual signals. Homing pigeons were particularly desirable because they had a good sense of direction, could fly high and their departure was unlikely to raise suspicion. They were officially in use through World War II, as shown in Figure 4-1. Encrypted messaging through telecommunication networks made the need for flying couriers obsolete.

Figure 4-1. Captain Caiger of the British Army Pigeon Service holding a carrier pigeon equipped with a "back carrier" message capsule during World War II. Source: Daily News and Getty Images-187949551.

Encryption has a long history of its own. In ancient times, When the population's literacy rate was low, writing itself was a form of encryption, because it was incomprehensible to most people, so the alphabet itself could be used as a code. But in general, a deliberate obfuscation of information normally has a secret algorithm or "cipher" with which to encode a message. The recipient needs to have a copy of the algorithm or a method, otherwise known as a "key," to decode or "decipher" the message. One such system used in wartime in ancient Greece was the Scytale code. It consisted of a long ribbon of parchment, wrapped edge-to-edge around a cylinder with a message written across it. When it was unwrapped, the message was not readable. The receiver of the message would have a cylinder of the same diameter to wrap the parchment around. In this case, the diameter of the cylinder was the secret key (Figure 4-2). The Scytale method is a form of what is more generally known as a "transposition cipher." The letters in the written message are transposed according to a rule, which is determined in this case by wrapping the parchment around the cylinder.

"Substitution cipher" works in a similar way by substituting every letter with another letter, combination of letters, or symbols. Substitution cypher is a superset of the transposition cipher and offers many more options. As an example, imagine if you touch type an English message on a Greek keyboard. You think you are typing an English word, but the outcome is an incomprehensible string of Greek letters.

Figure 4-2. The Scytale code used in Sparta and ancient Greece was a method of transposition cryptography. The unwrapped strip would not be readable.

By the middle ages, both substitution and transposition codes could be broken by analyzing the pattern of the message. For example, if the letter E is the most frequently used letter in English, then its corresponding letter in the coded message could be identified. To prevent this and hide any detectable pattern, substitutions could be changed dynamically. The so-called revolving code machines were invented to change the substitution in a complex pattern.

The encrypted message would have to be decrypted by a similar machine. The most famous example was the Enigma machine used by Germany in World War II.

In the twentieth century, coding machines were superseded by digital processors, as any complicated coding action could be emulated by a computer algorithm. To generalize this, every message can be thought of as a long series of numbers transmitted through a digital communication system.[33] The raw numbers are binary digits that are only zeros and ones. This includes any data, voice, or image used in various methods of communication. For example, a typical page of text can be transmitted by a string of approximately 150,000 binary digits. Pictures and videos are orders of magnitude larger, but still only numbers. These numbers can be encrypted prior to transmission and decrypted afterward. The cipher may be any reversible mathematical operation such as multiplication by a number on the transmit side and dividing by the same number on the receive side. Securing the messages this way slows down the communication speed, depending on the complication of the code. Typically, coding operations that afford higher security tend to consume significant additional encryption/decryption time.

There has always been a race between coders and code breakers, and typically, coders have a timing advantage over code breakers. Breaking a code may be cumbersome, while it is easy for the coder to just change the code to stay ahead. Of course, when the code changes, the new deciphering instructions must be sent to those who need to read the coded message. Code breakers would much rather intercept the key when it is being transmitted to one of its targeted recipients. The transmission of the decoding key to the receiver without interception, when the code changed, was always a major challenge for the coders and a weakness of any coding method. This weakness came from the fact that for centuries, cryptography was always symmetric: the same key used for coding was also used for decoding.

---

[33] For example, the letter E in ASCII Binary Code is represented by 01000101 which is number 69 in decimal notation. ASCII stands for American Standard Code for Information Interchange.

The revolutionary method of asymmetric cryptography was introduced in the 1970s. In this method, the encryption key and the decryption key don't have to be the same, so the encryption key can be made public, while the recipients use a private key for deciphering. This is also known as "public key cryptography." Of course, the decryption information must be imbedded in the public key as well, but retrieving it is very difficult. For comparison, let's consider encrypting someone's name by using his telephone number instead (Figure 4-3). A few decades ago, the standard method for doing this was to use a printed telephone directory.[34] A directory is analogous to a "public key." It was easy to look up someone's name in alphabetical order and find the associated telephone number (encryption). However, reversing the process was difficult. If someone intercepted a telephone number in a message, he would find it almost impossible to use the same public key directory to decrypt the message and retrieve the person's name, despite the fact that the information was contained within the public key. (The directory could not be searched electronically at the time.) The private key in this case would be a reverse directory in which the telephone numbers were listed in numerical order.

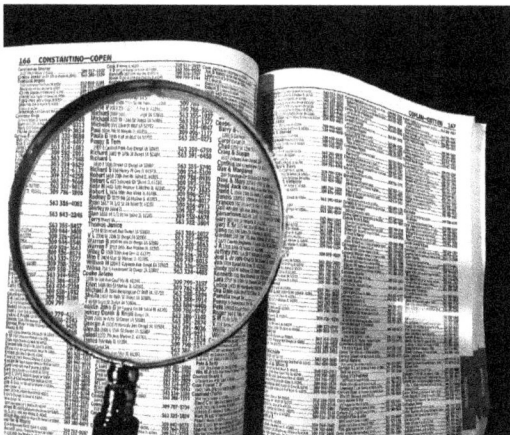

Figure 4-3. Public key cryptography is analogous to encrypting a person's name by replacing it with a telephone number. The phone book is the public key. Source: Jeff Cook, Quad-City Times.

---

[34] A telephone directory alphabetically listed the name of everybody who lived in a geographic region and provided the telephone number associated with each name.

Asymmetric cryptography takes advantage of the properties of large numbers. If the telephone directory had only a few tens of names it would not be a very effective public key. Its effectiveness came from the large number of entries in it. Since public key cryptography is computationally intensive, it is often used for renewing the key in simpler symmetric cryptography messaging.

Even though there are public applications of cryptography, such as digital signatures and cryptocurrencies, the most serious users are intelligence agencies. Today, every country has one or more intelligence agency with great capacity for gathering information not only about foreign adversaries, but also about every targeted individual in the society. This ability is steadily increasing with advancements in digital communication, ubiquity of surveillance cameras, social media, face recognition algorithms, and more. It is expected that in the near future, surveillance camera systems will be able to recognize almost any unmasked individual in their imaging range. You will be greeted automatically upon entering a store, for example, and your identity will be confirmed by wireless emissions from the gadgets you carry. At the same time, your location information and your daily activities will be accessible by law enforcement and intelligence organizations. Of course, processing this vast amount of information and initiating actions based on the processed information are very challenging, and methods for addressing this challenge are currently under intense development.

The difficulty in processing a large amount of information is related to the fact that most video and audio information is not "structured." In general, any information collected for analysis may be categorized as either "structured" or "unstructured" data. The difference between them has to do with our current ability to catalogue and cross reference each item. Structured data fits into a database, which is a large table with entries in numerous rows and columns. It allows data to be searched and categorized, as shown in Figure 4-4. The type of data that would be included in the database, for example, could be a recorded history of customer interactions in a commercial organization. This information may be searched by name of the customer, by date, or by type of interaction, among other criteria.

Name	Date	Item Purchased	Price Paid
John Smith	Dec 11, 2019	Tooth Paste	$2.50
Ali Milani	Jan 18, 2020	Picture Frame	$96.00
Kathy James	Feb 24, 2020	Jacket	$118.00

(a)

(b)

Figure 4-4. Examples of structured and structured data: (a) any data that can be tabulated is structured. (b) audio files and video files without any tagging are unstructured.

Unstructured data on the other hand includes sound, images, and video transmitted over communication networks that have no tagging or identification of content. This type of data is difficult to search and cross reference. The field of "Big Data" analysis is largely concerned with making use of the vast amount of unstructured data that is produced every day at an accelerating pace (Figure 4-5). Consider an indiscriminate video recording of automobile traffic at a busy intersection. The collected information is not of much use unless one can extract from the footage the license plate number of each car and perhaps other information such as the car's size, color, model, etc. This will create structured data that can then be searched and investigated to find out what time a particular vehicle went through the intersection or violated the traffic light, for example. With improvements in image and voice recognition, the goal is to convert most if not all such collected information into structured data suitable for searching and analyzing. For instance, a future reasonable query of a motion picture video file could be: "Show the scenes where actor A and actress B are having coffee." This type of search is currently not feasible.

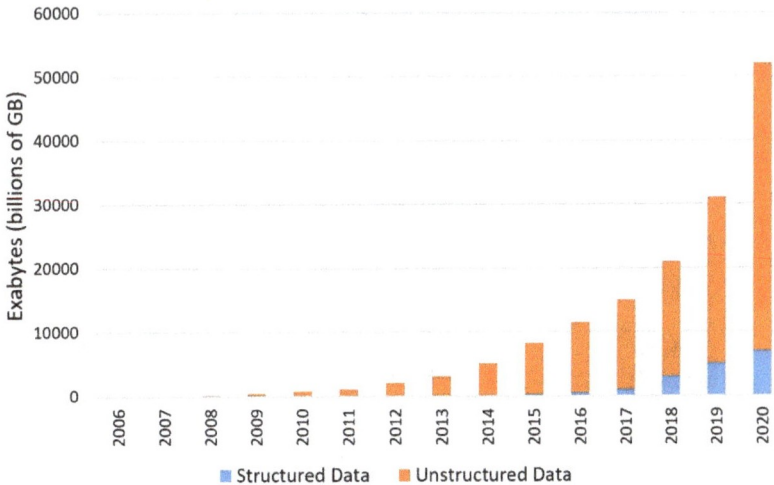

Figure 4-5. Data generation is predicted to double every two years for at least the next ten years. It is dominated by what is currently known as unstructured data.[35]

Throughout the late twentieth century, the vastness of a government's accumulation of data about the population favored individual privacy. For example, any single telephone conversation could be tapped and listened to, but monitoring every telephone conversation and extracting meaningful information would be very expensive and time-consuming. In other words, even though collecting, storing, and transmitting information became easier by the day, the volume of collected information surpassed the capability of any human or human groups to process and analyze. This data congestion precipitated a certain degree of privacy for the general population, as an individual's data was lost in the flood of collected information. Big Data analysis is about to

> Data congestion precipitated a certain degree of privacy for the general population, as an individual's data was lost in the flood of collected information.

35 EETimes, "Digital Data Storage Is Undergoing Mind-Boggling Growth," EETimes, September 14, 2016.

change that. Individual privacy is being eroded by algorithms that are becoming more effective in tagging and formatting unstructured data.

Humans are inefficient in analyzing raw data, such as telephone conversations or video footage, but they are still quite effective in the final analysis of processed data. Suppose the facial images of every person attending a sports event in a stadium is captured and the identification of a single individual is desired. A human cannot process that amount of raw data, but if a computer algorithm reduces the number of potential targets, to say, below fifty, a human or a group of humans can very reliably reduce the set to one final target.

In the long run, because of the lengthiness of the process, this final analysis is also not appropriate for humans. If the identification of the target is for security purposes, for instance, the time it takes a human panel to make the final identification may be too long, leaving little time to intercept any threat. Therefore, as time goes on, humans will have to be removed from the processing and analysis of this type of information altogether. Big Data analysis will help us achieve this goal. When all data is automatically processed, no one will be able to seek concealment anymore by hiding behind the volume of unexamined information.

J. Edgar Hoover, the first head of the FBI (Federal Bureau of Investigation) in the United States, started his career as a library clerk.[36] In those days, every library had catalog cards kept in drawers, which allowed patrons to search for a book and locate it on the book shelves according to its title, author, or subject (Figure 4-6). As the head of FBI, Hoover was quoted as saying that FBI should keep similar library catalog cards on everyone living in the United States[37]. Presently, the cataloging by security agencies may not include every single person living in the country, but it does cover the vast majority

---

[36] Richard Gid Powers, *Secrecy and Power the Life of J. Edgar Hoover* (London: Arrow Books, 1989).

[37] Alana Mohamed, "How J. Edgar Hoover Used the Power of Libraries for Evil," Literar Hub, March 5, 2020.

of the population. Also, the rapidly rising amount of information kept on every individual is far greater than what was originally on the library-style catalog card envisioned by Hoover.

Elements of a Catalog Card

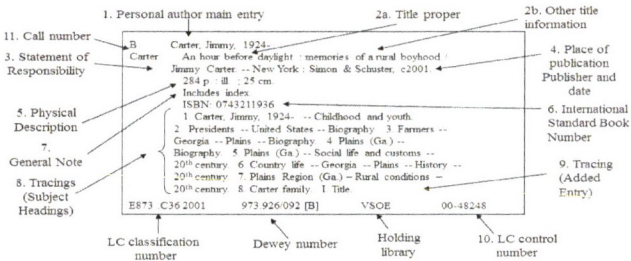

Figure 4-6. A library card catalog and the information listed on a card (library.net).

Currently, the intelligence held in the vaults of each government is not limited to its own people and institutions. Information is kept in every country on every individual, place and relationship they have access to. This information is often shared with friendly governments. During the Cold War, groups of countries formed intelligence-sharing blocs. These blocs have changed over time, but the general structure has persisted to the present day. In today's world affairs, allowing or not allowing external intelligence access often determines if a country is tagged as a friend or a foe. From this perspective, there are only a few extended intelligence domains in the world that operate independently of each other.

The old adage "knowledge is power" raises an intriguing question: Is there a way to regulate and supervise information collection by intelligence agencies? Is the legislature or the executive branch of the government capable of passing regulation that goes against the will of an intelligence agency? Is it possible for the government to maintain control by funding and commissioning multiple intelligence organizations? What if such "competing" organizations have shared interests and choose to cooperate instead? Isn't it possible for an intelligence agency to make itself immune from external checks and regulation?

Any answer to these questions must consider the fact that publicized private information can make or break politicians.[38] Therefore, it is difficult for a politician to be successful in controlling and regulating intelligence organizations without their cooperation. Controlling their funding is also not an effective method, as funding available to an intelligence agency is not limited to its official government budget. They are capable of harnessing income from various hidden sources such as arms sales and drug trade. One such violation of the government's stated policy by an intelligence agency was revealed to the U.S. public in the 1980's.[39] In that case, revenues from one covert activity were used to fund another in a different region of the world without any official accounting of the funds. In addition to being a source of income and independence, an intelligence agency's involvement in hidden transactions benefits them in many other ways. For instance, individuals identified by the agency in a drug-related transaction can be blackmailed into cooperation for other purposes when needed.

Intelligence services control and protect the population in ways that are not necessarily discoverable by the general public. Security threats are neutralized quickly and quietly before they reach the headlines. In addition, intelligence agencies have recently become involved in

---

[38] Frontrunner U.S. presidential candidate Gary Hart (1988) had to withdraw his candidacy once compromising private information about him became public.

[39] Iran-Contra Affair, *Encyclopedia Britannica*: www.britannica.com

collecting economic and environmental information that is useful in international negotiations and infrastructure planning.

Maintaining an information database on people's routine activities is growing despite its controversial nature. People voluntarily provide information about their personal contacts and daily interactions to social media that naturally maintain such databases. Some governments are not denying their attempts to closely track the everyday activities of individuals. With the continuation of this trend, information-gathering organizations may be starting to form one of Simorgh's key body functions analogous to a nervous system that gathers information and sends it to one or a few central locations for processing or storage. It is essential for Simorgh to collect real-time information on all its human nodes[40] in order to maintain harmony and address the needs of each individual. This process becomes less threatening when we realize that the bulk of data collection and sorting is carried out automatically and without human interference. Even the analysis of the information and any subsequent action will soon be carried out at command hubs without human decision makers. Nonetheless, many people perceive a loss of privacy as a threat.

> It is essential for Simorgh to collect real-time information on all its human nodes in order to maintain harmony and address the needs of each individual.

Privacy

Our private behavior changes as soon as we suspect that we are being watched. We have privacy if we are free of any such suspicion. The same concept applies to groups of people such as families and clubs with members who have shared interests. They also behave differently as a group when they are watched.[41] Historically in the

---

[40] Each human node of Simorgh is a person who is comfortable with being immersed in high technology. See Chapter 15.

[41] "Intrusion into seclusion" with some exceptions is forbidden by law in many countries. Its degree of invasiveness is determined by how the information is misused.

wild, it was a sound or a motion that aroused the suspicion of being watched. This would switch the person's private behavior to social behavior. The triggered emotion could be fear (of a lurking predator), anxiety (of losing one's food or treasured belongings to an intruder), indifference (being watched by a harmless creature), joy (of having company), or eagerness (to meet a potential mating partner). Being watched in today's digital age can trigger a similar range of emotions. Numerous people willingly post their most personal information online, hoping that those who pay attention are neither predators nor thieves. A predator could use such information to inflict bodily injury, while a thief could impersonate someone and gain financial and legal access to his or her possessions.

When the emotions of fear and anxiety associated with loss of privacy linger for a long time, most people become passive and acquiescent. People in power can take advantage of this to intimidate and pacify the population. This is potentially the most damaging side effect of the loss of privacy in modern times, and works against the argument that only the guilty should be concerned about privacy. If you have "nothing to hide," you still need to be weary of having your personal information exposed to predators and thieves, including a corrupt and non-representative government who could punish you for dissent.

Private information has monetary value. As the role of technology becomes more prominent in our lives, we see that our private information is the currency with which we buy convenience. If you don't want to reveal your shopping habits, you will not enjoy the convenience of cashless transactions. If you don't want your location to be tracked, you will not have the convenience of receiving directions to your destination.

Providers of goods and services need to know people's shopping preferences and lifestyles. They use this information to tailor their offerings to the needs of their customers. This is not a new concept. A local village grocer always knew the consumption patterns of his customers and managed his inventory accordingly. Now that sellers to the global village are not able to probe their customers' preferences directly, they purchase the information from intermediaries such as Google, Apple, and Facebook, among others. This also allows them to do targeted advertising to improve sales. The involvement of intermediaries places a monetary value on people's private information. Each person influences the value of this information by deciding how much convenience he or she is willing to exchange for it.

Much of the concern about loss of privacy stems from the possibility of personal information falling into the hands of the wrong people (predators and thieves). Otherwise, the majority of the population does not appear to have any trepidation about private information being securely stored in a bank's computer or in a hospital. Another way of looking at this is that known and predictable uses of private information are acceptable. It is concerning when we do not know who will receive the information and how it will be used.

In the process of closely monitoring its nodes, Simorgh will generate and store a significant amount of "private" information. The collection of this information by Simorgh is akin to current data storage at a hospital or a bank even though the volume of information will be much more extensive. It is essential, however, that this information does not leak to outsider humans and other Simorghs who may use it with malicious intent. Simorgh will have defenses to safeguard any such database against potential leaks and attacks.

The government or the central command of Simorgh will not be power grabbers who would be inclined to suppress the population. The central command will be clusters of nodes guided by intelligent algorithms written to harmonize the functions of Simorgh. Unlike with human decision makers, these algorithms are transparent and free of deception. Any built-in biases are discoverable and may be iteratively refined over time. Eventually, they will be free from any partialities or provisions that could inadvertently discriminate against any node. Since no individual rights will be violated, privacy will not be required as a tool to protect the population from central command infringement. Nevertheless, there are personal secrets in an intimate sphere of privacy that most people prefer not to divulge and Simorgh will have no interest in sensing and recording. The privacy of such information is expected to persist indefinitely.

Summary

Extensive real-time information about the population is required for the government to maintain power and control, as well as to provide services when and where they are needed. Information about the foreign population is also needed for security and defense purposes. Information networks of any country or a group of cooperating nations has the potential to evolve as Simorgh's central nervous system. Simorgh's human nodes will need to give up some privacy. The scale will be similar to the level of privacy that people are already relinquishing by using social media and other electronic services that make life more convenient. By construction, Simorgh will use the data collected on individuals to address the needs of its human nodes. Safeguards must be put in place to prevent the leakage of such information to outsider humans or hostile Simorghs.

Chapter 5: History of Human Communication and Interconnectedness

Simple communication between animals began and continues to be in the form of gestures and sounds to address nearby individuals or groups. Gestures reflecting emotions are common among almost all mammals and primates, and cover basic emotions such as affection, rage, happiness, submission, etc. Communication by sound not only supplements gestures, but can carry information over longer distances beyond the originator's visible range. A crying human baby can be heard a few street blocks away. Howler monkeys can be heard a few kilometers away. The range of sound can be much longer in water, with whale songs having the farthest reach of up to thousands of kilometers.

Early communication among humans took a quantum leap with the development of language. Objects, individuals, and actions could be named and thereby uniquely distinguished. This created an exceptional capability for humans to learn through the experiences of others without direct observation. Language was the first step toward building what we know as "culture." Written language made it possible to extend learning to successive generations without direct apprenticeship.[42] This luxury is exclusive to humans, and it is a prerequisite for advanced learning and progress. For centuries, reading and writing was practiced solely by a select group of scholars and scribes. Levels of literacy in society typically did not exceed ten percent.[43] Books were rare, and most people were exposed to written material through listening to its recital by literate people.

Figure 5-1. A scribe in ancient Egypt in charge of reading and writing in the King's court. From Canadian Museum of History (www. historymuseum.ca).

When it was written on a hard tablet, single-copy text survived for a very long time. Many such tablets have been recovered from archeological sites. Text written on soft material such as animal skin or papyrus had less of a survival chance, and the majority of such early manuscripts have been lost forever. Religious books, and texts written by famous authors were copied by students and scholars manually and kept for widespread access in multiple libraries.

Written long-distance communication was made possible by courier services that took the material such as a royal decree to various provinces of a kingdom. Such courier services later became

---

[42] Written language was originally a sequence of pictures that got simplified into characters. Phonetic writing can be seen in the evolution of the cuneiforms initiated by the Sumerians. The alphabetic system developed by the Phoenicians (1,500 to 300 BC), became the basis of most phonetic alphabets in use today.

[43] William V. Harris, Ancient Literacy (New York, NY: ACLS History E-Book Project, 2004).

formalized as postal services to carry mail to distant recipients.[44] In this way, the postal service created the infrastructure for both "broadcast" type communication, as in formal announcements to everyone, and "point-to-point" communication in the form of personal letters.

The major milestone after which written information became widely available to all was the advent of the printing press, attributed to Johannes Gutenberg of the fifteenth century. Since then, printed materials (books, and later, magazines and newspapers) became the main broadcast form of communication, while letters and mail service remained the means of point-to-point long-distance communication.

The advent of electricity brought about the next revolution in communication. The telegraph (Figure 5-2) became the first vehicle of instant remote messaging.[45] It gave a boost to international commerce and brought the world a lot closer together. An international business transaction that used to take weeks or months to complete through the mail system could now be concluded within a day or two using the telegraph.

Figure 5-2. An early telegraph switch and the Morse code representing the distress signal "SOS" (Courtesy of the Smithsonian Institution). The Morse code for SOS is shown on the bottom left.

---

[44] The informal motto of the US postal service, "Neither snow nor rain nor heat nor gloom of night stays these couriers from the swift completion of their appointed rounds" is a translation of what was originally written about the ancient courier service "angarium" around 400BC.

[45] Samuel Morse was instrumental in the development of the telegraph in the early 19th century.

The telegraph was capable of transmitting information at the rate of approximately five bits per second.[46] It used "Morse Code" which utilized pulses with two different durations. Each transmitted bit was either a long pulse (dash) or a short pulse (dot). Each alphabetic letter was represented by the combination of a few such bits (Figure 5-2). For example, the letter S was represented by three dots, and the letter O was represented by three dashes. Signaling was initially done by hand, and each word could take a few seconds to transmit. The telegraph was soon superseded by other forms of telecommunication, such as the telephone, radio, and television, each providing a higher "bandwidth" or information content. For comparison, a telephone conversation is transmitted at the rate of approximately twenty thousand bits per second (20 kbps), and a 4K television signal is transmitted at approximately sixteen million bits per second (16 Mbps). It would take over seven hundred years for an old telegraph line to transmit the information content of a 2-hour 4K movie.

Digital computers were initially introduced as number-crunching machines that could compute multiple and repetitive mathematical operations; hence, the name "computer." They replaced the mechanical calculators used primarily by accountants (Figure 5-3), as well as the slide rules and the lookup tables that were used by engineers. It was not easy to interact with early digital machines, as the system could only understand inputs in restricted digital formats. A card punch machine is an example shown in Figure 5-4. This machine would transcribe each typed line of the program into a card with punched holes in it, to be fed into the computer. A computer program that the user wanted to run would often be a large stack of such cards.

---

[46] In modern digital communication, a bit of information is represented by either digit zero or one. It may be thought of as a yes or no answer to a question. The usefulness of the bit of information depends on the question that it answers. But the capacity of a transmission channel is limited by the number of bits per second that it is capable of sending or receiving.

Figure 5-3. Mechanical calculators: Pascaline of the 17th Century (left)[47], and a more refined unit used in the nineteenth Century (right).[48]

A keyboard and display screen that were directly connected to the computer emerged later as a more user-friendly way to interact with the machine. Voice recognition and natural language communication are still evolving. At first, people had to learn the language of the computer. Later, a paradigm shift enabled by the increasing sophistication of technology required the machine to learn human language instead. The holy grail is a direct human neural interface to the computer (Brain Computer Interface, BCI) that is bound to become available in the near future.

> The holy grail is a direct human neural interface to the computer that is bound to happen in the near future.

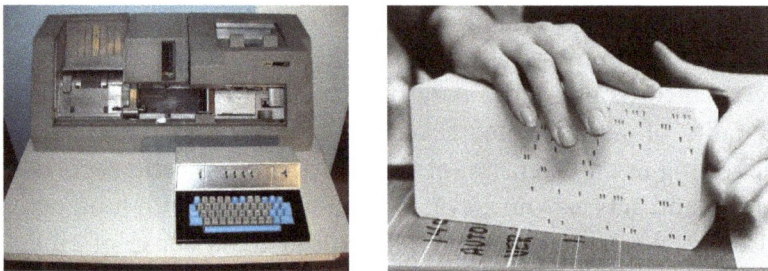

Figure 5-4. IBM-29 card punch machine (left)[49] and a stack of punch cards prepared to input into the computer.[50]

---

[47] "Pascal's Calculator," Wikipedia (Wikimedia Foundation, March 9, 2021).

[48] Monroe LA7-60, Monroe Calculating Machine Company BV., Amsterdam, Netherlands.

[49] "Keypunch," Wikipedia (Wikimedia Foundation, March 31, 2021).

[50] Bill Buchanan, "IBM's Greatest Challenge?," Medium (October 12, 2020).

Early computer networks were developed to facilitate data sharing among work groups in universities or within companies. Initial networking procedures were written for functionality but not widespread compatibility. Diverse networking protocols started to converge following a U.S. government initiative in the late-1960s that led to the protocol known as Arpanet. (ARPA is the U.S. government's Advanced Research Projects Agency). Arpanet implemented the new concept of sending information in small bundles, called packets, that were independently transmitted and then recombined at their destination. This concept of packet switching and its successors replaced the more restrictive "circuit switching" and enabled larger computer networks to form. Finally, a network of networks emerged known as the internet, to which most computers are connected today. This final unification and worldwide acceptance of the Internet Protocol in late-twentieth century was a major milestone in the evolution of human communication. It is now bringing libraries of information to everyone's fingertips.

Once mobile phones were connected to the internet, they transmuted and became the key personal hubs for human networking around the world. Today's personal wireless phone has vastly superior capabilities over the room-sized computers that started the original data networking process. In addition to sending and receiving text, voice, and video messages, a "smartphone" now can act as a sensor when it records biometric data such as heart rate, physical activity, or quality of sleep. It may act as an actuator and remote control, functioning as an extension of your arm. It senses and reports your location and acts as your navigation tool. It helps with food delivery and entertainment. You can even experience a sense of emersion in a virtual environment projected on the mobile phone's screen by wearing a special headset and perhaps a pair of gloves that will allow you to manipulate the virtual environment (Figure 5-5).

Figure 5-5. With a headset and gloves, you can see and feel a remote virtual environment (From Shutterstock).

The current development of the "Internet of Things" or IoT is another step in the direction of connectivity enhancement. It brings a wide variety of objects, machines, appliances, humans, pets, etc. and gives them direct access to the network. Each item has the ability to transmit and receive data in real time. An example is home-automation, which allows the user to monitor the environment of the house and take action remotely when needed. Another example is continuous monitoring of the vital signs of individuals for personalized medicine or for automatically reporting a personal emergency.

From a more general perspective, IoT adds sensing and motor-function capabilities to the internet. Mobile phones, wearables, and vehicle-mounted sensors transmit information on the go. Other functions such as home automation and environmental sensing employ mostly stationary sensors. Remote activation of garage doors and remote operation of robots and machines provide motor functions. Sensor additions to the internet make it possible to know about environmental conditions, traffic flow, and energy use, in addition to location, health,

> Individuals who choose to be employed within Simorgh will find themselves under full-time observation and care.

and emotional status of individuals. Motor functions provide the additional capability of reacting to such conditions as needed.

Eventually, the cellular phone will evolve to be each individual's permanent connection to Simorgh's central nervous system, which will be the future of the internet. Individuals who choose to be employed within Simorgh, will find themselves under full-time observation and care through stationary and mobile sensors, including their own permanent connections to the nervous system. At the same time, the system will provide for all their material, entertainment, and survival needs. As a body node of the larger organism, every fully employed human will have the responsibility to receive information related to his occupation and to transmit processed information and commands on the basis of his job function and specialization. Plenty of extra time should be available for leisure and entertainment. Eventually, for those who become accustomed to this comfortable life style, functioning and survival without being connected in this manner will become difficult or even impossible.

It is conceivable that work instructions for each position in a company, or in the society as a whole, will be stored at a central location, accessible to all, but filtered to show the information relevant to the position of the inquirer. This is already practiced in many companies through what is known as "enterprise software." It is reminiscent of the information stored in the DNA molecule that is accessible to all cells, yet each cell accesses only a small portion of it as needed.

Summary

People are becoming more closely linked to the network that is poised to become the principal nervous system of Simorgh. Internet of Things is enhancing the sensing capabilities and adding motor functions that activate Simorgh's "limbs," allowing it to engage in numerous simultaneous mechanical projects such as construction, mining, and exploring. Point-to-point and broadcast-type communication systems will emulate the nervous and endocrine systems of the society. Algorithms and stored information will provide all the necessary daily work instructions to each human or nonhuman node. This is akin to a DNA code for Simorgh.

Chapter 6. Information and Truth

Each person's mental image of the world is unique and private. Our mind creates for us a picture of the world through which we can interact more easily with our environment[51]. The attributes of this image are shaped by the tools with which we probe the environment and by the subsequent processing of the information that reaches our minds. For example, probing our surroundings only with our hands, gives a different mental image than probing with the light that reflects from an object and reaches our eyes (Appendix C). Our mental images define what we perceive as reality, which does not have to be the same as another person's perception (Figure 6-1).

There is a biological condition known as Synesthesia[52] that affects about four percent of the population and causes the double perception of colors, sounds, tastes, etc. Synesthetes may experience tastes that they associate with colors, or lights and colors associated with musical notes. These associations are consistent and not imaginary. Many synesthetes become aware of their uncommon perceptions only by chance when they happen to compare their experiences with others. It is important to note that it is the double perception

---

[51] Donald D. Hoffman, "Sensory Experiences as Cryptic Symbols of a Multimodal User Interface," Activitas Nervosa Superior 52, no. 3-4 (2010): pp. 95-104.

[52] S. Scutti, "What It's Like To Experience Synesthesia: The Taste Of Music And Colors Of Language", Medical Daily, Mar 7, 2014.

that makes this condition detectable; otherwise, there is no way of knowing if someone perceives colors the way you perceive sounds, or if your perceptions have any similarity to those of others. From childhood, when you are shown the color brown for example, you are taught to call it brown regardless of the sensation that it generates in your brain. It is only the consistency of the nomenclature that allows you to engage in a conversation with another person or even a computer that has learned to correctly identify and name various characteristics of objects. It is irrelevant whether others use a mental image like yours for the identification.

> It is only the consistency of the nomenclature that allows you to engage in a conversation with another person or even a computer.

By the same token, you cannot expect any similarity between your mental images and the outside world. Even if you assume that the reality of the physical world is unique and independent of an observer, you only deal with icons that your brain generates to represent the outside world. This is not unlike the relationship between an icon on your computer screen and the actual software program that it represents. The correlation is one-to-one, but there is no physical resemblance. The most you can expect from your mental images is consistency.

Let's look briefly at the epistemology of a few words relevant to our discussion. These words are truth, reality, opinion, and belief. To start, there is a subtle difference between truth and reality: truth may be thought of as the validity of a statement. For example, the statement, "the dog bit my brother's leg," can either be true or false. Reality, on the other hand, is the plurality of what surrounds the truth and is not limited to one fact. It may be related to the ownership of the dog, the type of dog, the circumstances leading to the bite, and so on. Another example is the illusion of a magician's act and the reality that is behind it (Figure 6-2). Based on the truth and reality that are presented to us, and in conjunction with our own thoughts, we may form an opinion or a belief. Of course, a belief commands a much higher level of certitude than an opinion and usually influences our choice of lifestyle.

Figure 6-1. Self-perception is typically very different from how others see us. Is reality unique?[53]

We often hear the expression "the whole truth".[54] In a legal context, this refers to not misleading others by offering selected parts of a story. In the strict sense of the expression, the "whole truth" about an event or a system is impossible to know because it contains too much detail. For example, in the dog bite scenario, the whole truth would involve the life history and the habits of the dog, the temperature and humidity of the room, the objects present in the room, the smells in the environment, and so on. The story that we tell about the dog bite cannot be the whole truth. Our knowledge is always incomplete and therefore, it always entails a high degree of subjectivity.

In a more abstract sense, the reality of the world around us has many different scales, from macroscopic to microscopic[55], with numerous mutual influences and correlations. Therefore, depending on our focus, we routinely miss or ignore a large portion of the reality that we are immersed in. For humans in general, being able to tell a self-consistent story is more desirable than striving for the whole truth. We tend to ignore the details that are not relevant to the story.

---

[53] M. Scoda, "Perception: Do You See What I See", mariascoda.com.au, December 2014.

[54] "The Whole Truth" was a research topic in Philosophy at the University of Glasgow in 2016 and 2017.

[55] From microscopic to cosmic scales, there are layers of reality. Each layer's details may or may not be significant to the next layer. More detail is provided in Appendix C.

Since it is the story or the narrative that matters, reality doesn't have to be unique, even as perceived by a single individual. It is possible for one individual to live in differently perceived realities at different times. One may experience different realities at work versus at home, depending on the work environment. This duality may be as extreme as having different beliefs and opinions while at the two locations. This is possible if one focuses on different aspects of reality at each location.

For example, if you are an astronaut, you may see Earth as a beautiful blue sphere when you are on a mission. While at home, you perceive Earth as the flat land on which you walk. An earthlier example is a peculiar phenomenon that some people experience during sleep. It is called "continued dreams" where a sleeper experiences a story line and sequels to each dream from one night to the next.[56] This may go on for days, weeks, or months. It can feel like you are living in two different worlds. Sometimes, your value systems and social status are different in the continued dream versus real life. Another example may be found in computer games. As a player, you are often plunged into a fantasy world that you choose to believe or adapt to temporarily while playing the game. This concept of immersion into computer games can be extended to the future of work where you may be immersed into an artificial reality or a simulated world that helps you perform your job functions better (Chapters 7 and 8). In general, people don't mind switching between realities, as long as in each reality there is a consistent story to tell.

Figure 6-2. Illusion vs reality. The two center squares are the same color in isolation but perceived differently in the context.[57]

---

[56] P. McNamara, "Why Some of Your Dreams Have Sequels", Psychology Today, December 30, 2014.

[57] "Optische Illusie - Over Fysiologische Illusies En Schijnwaarnemingen," Pieter Broertjes, January 20, 2020.

Memories play a large role in everyone's individual perception of reality. Unlike electronic memory that can preserve words, speech, and images in a faithful manner, human memory is not very reliable. Referring to your recent memory, you can easily

> An event is not stored in your memory as a continuous movie, but rather in snapshots that you combine at the time of recollection to reconstruct a coherent story.

recall most of the activities you were involved in yesterday; however, if you try to reconstruct every detail, you will realize that the events were not stored in your memory as a continuous movie, but rather in snapshots that you combine at the time of recollection to reconstruct a coherent story. You may remember that you went to a building and were greeted by someone at the door, but it is unlikely that you remember the shapes and number of stairs that led to the building. If you are asked to remember them, you may fill in the missing detail by guessing or from previous experience if you frequently visit that building.

The number of stored snapshots in your memory decreases the farther back you go in time, and it takes more of a mental interpolation to reconstruct the story. Each time you recollect an old event, you create and add new snapshots to fill in the gaps, and these new snapshots are added to your original memory of the occasion. In other words, the act of recollection by itself can modify your memory of an event. Over time, your mental bias tends to alter the recalled events, and the tale leans toward what you prefer to remember. There are many documented records of childhood memories of events that never happened[58]. In criminal justice, there are occasions of victims positively identifying the wrong suspect. You may have experienced similar inaccuracies in your memory when you and your partner have conflicting recollections of an old event at which you were both present. Photo albums and recorded movies help us replenish our distant memories by giving us a chance to insert more accurate

---

[58] Lampinen et. al. (1997, September 1). "Memory illusions and consciousness: Examining the phenomenology of true and false memories." *Current Psychology*. 16 (3-4): 181-224.

snapshots back into the slide show of the stories that we remember.

We should not be disturbed by the fact that in everyday life the realities we perceive are not unique nor the accounts of daily events we remember necessarily accurate. Accuracy is not needed beyond what is useful in our daily lives.

History of Information Sharing

Early humans depended solely on their own five senses (touch, sight, hearing, smell, and taste) for collecting information about their surroundings. Their views and convictions were survival-oriented. They needed information to decide where to shelter, how to tell a friend from a foe, and where to gather food. Their own experience or observations of success and failure in the wild, helped them derive personal plans for future use. The information had no external bias because it was all self-collected.

Spoken and written language, as well as travel, made it possible for people to receive indirect knowledge obtained by other people. Information collected by others had an element of bias and subjectivity because it was conveyed from the point of view of the messenger. In some cases, the consumer of the news would question its validity and credibility. People tended to scrutinize messages that had a direct impact on their daily survival more than news that would shape their view of the world in a broader context. For example, the news about a predator nearby would be examined carefully, while news about the lifestyle of a gaudy chief in a faraway tribe would be taken at face value. This tendency to examine the validity of local news more closely has persisted to the present day.

Over the last few centuries, people in industrialized communities have relied heavily on newspapers and magazines for information about the world. Newspaper printing started in the seventeenth century in Europe and was initially heavily sanitized and controlled by the government. The popularity of this medium increased as government censorship relaxed in many countries, even though every government still maintained some control over mainstream news. High-speed press became available by the mid eighteenth century and enabled the mass production of newspapers at a very low cost (Figure 6-3).

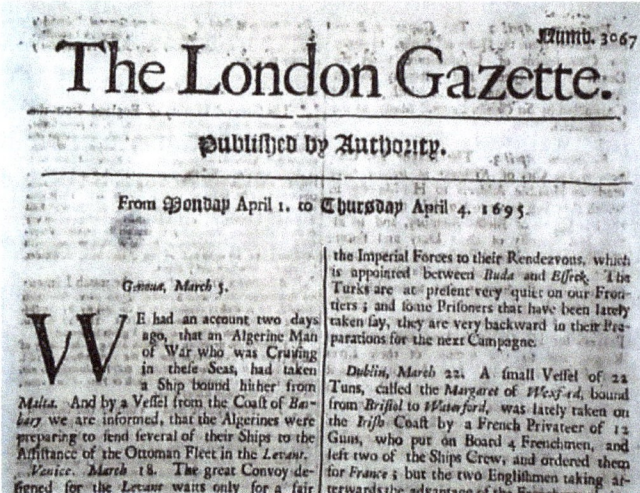

Figure 6-3. The London Gazette is possibly the oldest English language newspaper still in publication. This issue is dated April 4, 1695.

Each newspaper targeted a specific audience. Politically, different newspapers supported different government factions. There were religious newspapers, economic newspapers, and those that covered gossip and unsubstantiated news. Readers could choose to limit their exposure to one or more of these types of newspapers to satisfy their information needs or strengthen their own points of view.

Another mass media outlet for news and entertainment that became popular in early twentieth century was the radio. This revolutionary wireless transmission of human voice grew rapidly in popularity. It was used for news, music, entertainment, and sports broadcasting. Its popularity grew further with the introduction of transistor technology which enabled portable radio receivers. Except in the West, radio stations in most countries were owned and operated by the government; consequently, most people received the "facts" about world affairs through the filter of their government. Shortwave radio made worldwide broadcasting possible and gave polyglot enthusiasts the chance to hear other versions of the news. Shortwave broadcasting was done with positive and negative purposes, ranging from well-intentioned news, entertainment, and educational material to propaganda and subversive agitation. Radio broadcasting has retained much of its popularity in part because it can "play in the background" at home, in the car, and in public places.

In the mid twentieth century, yet another medium, namely the television, began to dominate the dissemination of news and entertainment to the public (Figure 6-4). The television became the focal point of the average western household and brought family members together, particularly during early evening hours, or "prime time." At first, only a few major networks dominated the airways due to the relatively high cost of TV transmission stations and production of programming. The major networks provided a "standard" viewpoint on news, social issues, and world affairs. The centralization of newscasting in this way created a mainstream standard for national and international reporting. During this period, TV news was highly refined and far less sensationalized than in print media. Even though the interpretation of the news and the resulting opinions varied among individuals, the perceived facts and knowledge on which everyone's opinion was based, were the standard narratives seen on TV. This created a foundation for constructive political debate in countries where such free debate was allowed.

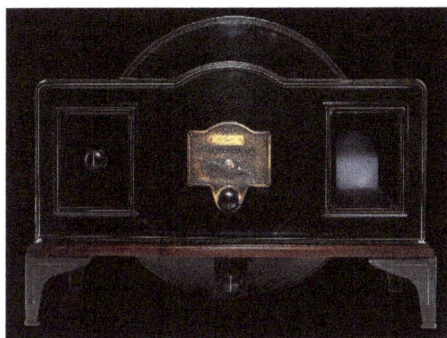

Figure 6-4. Baird's mechanical television that was debuted in England in 1929 had a small and blurry image. Next generation televisions used cathode ray tubes (CRT) for image formation, later replaced with flat panel liquid crystal display (LCD), and lately by LED and flexible OLED.[59]

[59] "Invention of the Television," DK Find Out!

Music, theater, nature, and sports were among the other categories of TV programming, as well as other entertainment and educational topics. People who browsed TV channels for subjects of personal interest often found programming that was not quite what they were searching for. They would thus be exposed not only to content that partially matched their interest, but also to programming that deviated somewhat from their original impulse. This resulted in exposure to different points of view and the spawning of new interests, thus broadening the viewers' perspective on the world.

TV viewing was similar to how subscribing to a newspaper or magazine facilitated unintended encounters with articles of social value, with the added benefit that such sightings on TV were often shared by the whole family. The cohesion that television brought to society and families was short lived. It began to fade as television lost its status as the principal provider of information and entertainment. In the twenty first century, both information and entertainment became available "on demand" through the internet in a very convenient and addictive way. A negative side effect of delivering snippets of information on demand was that it only exposed the audience to the material they searched for, without exposure to content they *didn't* seek. Inadvertently, this limits one's information horizon and encourages people to live in partial seclusion without being cognizant of it. Additionally, automatic filtering on the internet that only feeds content of interest to each individual reinforces this isolation and causes people to live in what is referred to as a "filter bubble." Interest in such intellectual isolation builds barriers and social intolerance. It fragments society and may incite a desire to obtain and/or produce false information in order to reinforce one's stance.

> Interest in intellectual isolation can often lead to a tendency to intentionally obtain and/or produce false information.

It is tempting to blame online social media for this isolation, as they facilitate the formation of bubbles within which people give and receive confirmation bias. However, one should note that ideological bubbles have always existed, even in the absence of selective information filtering. Most people enjoy socializing with friends and colleagues who think substantially alike. They may show no interest in hearing opinions and viewpoints that contradict the basis of their daily discourse. In other words, many people would prefer to live in their own comfort zones than in reality. Historical popularity of ideological and faith-based establishments, such as churches, clubs, and fraternities, confirms people's desire to stay with a pleasant familiar narrative than encounter the harsh reality of the world that they may find to be multifaceted and confusing.

The current upward trend in the formation of social and ideological bubbles is unlikely to reverse. In fact, technologies such as virtual reality, will allow people to live in a world of their own choice, and avoid the ups and downs of reality to a large extent.

Disinformation and propaganda

There is undoubtedly a sizable market for false information as evidenced by its sheer volume. Let us investigate the reasons by focusing on the disseminators and consumers of public information separately and identifying potential incentives that each may have to deviate from the truth.

Disseminators' reasons for reporting skewed, false, or misleading news may be categorized as follows:

1. Filtered and partial reporting: The reporter's own opinions always distort and filter the information to be reported. Depending on the conclusion that the journalist desires to reach, some observations are emphasized or exaggerated while others are overlooked. For example, during the Cold War, reporters from the Soviet Union tended to focus disproportionately on the plight of the homeless in the United States.

2. False advertising: False claims about products and services can attract business even though they are outlawed. For example,

the expression "this toothpaste whitens your teeth noticeably within a week" may convince many people to purchase it. The statement may even have an element of truth to it, but it is clearly designed to be misleading.

3. <u>Attracting audience:</u> Apart from advertisements, the news media itself, such as a magazine, a newspaper, or a website, usually receives considerable revenue from subscriptions, clicks, or hardcopy sales. There is a paying audience for sensational news or "yellow journalism" (Figure 6-5). For example, many people like to hear about alien visitors from outer space and the government's conspiracy to keep it a secret. Another group of people might be interested in celebrity gossip, and so on.

4. <u>Manufacturing consent and forming public opinion:</u> False claims about a potential enemy can provoke war sentiment. Falsely embellished claims about a bill that is in the legislature can attract popular support or popular indignation. In many cases, there is an indirect economic incentive for manufacturing consent.

5. <u>Supporting an Ideology:</u> False news in support of a school of thought or against its critics may only have an ideological benefit and possibly little or no economic value. False news by "pro-choice" or "pro-life" activists about the abortion debate in the U.S. is a noted example.

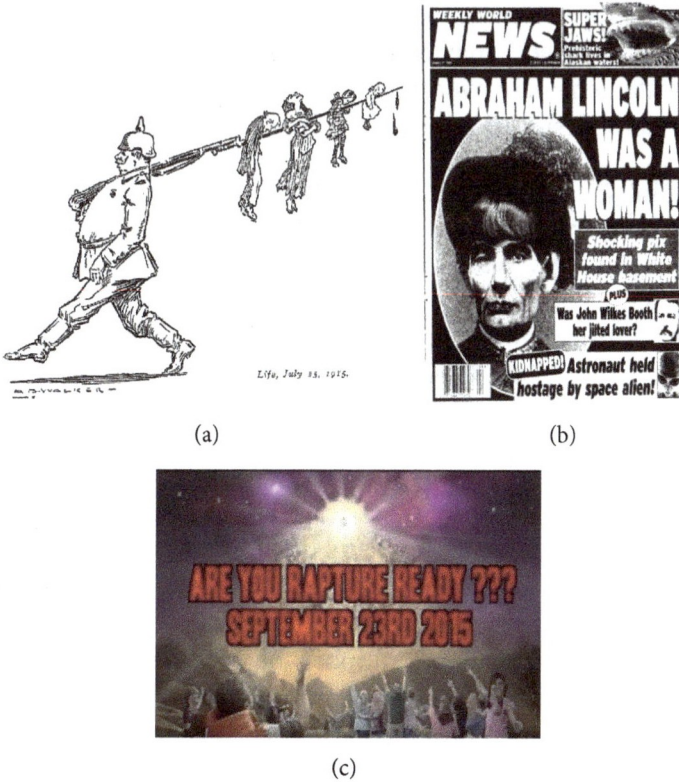

(a)

(b)

(c)

Figure 6-5. Examples of Disinformation: (a)[60] Manufacturing Consent in WWII, showing a German soldier impaling the elderly, women, children, and babies (b)[61] Yellow Journalism using shocking headlines to capture attention, and (c) Ideological Propaganda.

Unlike the sources of false information who tend to knowingly promote it, the recipients may or may not be aware of its falsehood. In any case, the following aspects of false information have been found to be appealing to different recipients:

1. Gossip

   Gossip is a way for humans to bond in pairs or within a group. Sharing dislikes about others creates an enjoyable conversation

---

[60] M.B. Walker, "German Bayonetting Children," Life, July 25, 1915.

[61] "Abraham Lincoln Was a Woman," Weekly World News, January 1, 2002.

and strengthens the bond within the group. The joy comes in part from implying moral superiority over others, and in part from drawing vicarious pleasure. If you gossip about someone's promiscuity, you imply that your character is superior; at the same time, you may enjoy imagining the experience as being your own. Gossip often contains traces of truth, but repeated gossip tends to be exaggerated and over inflated. Typically, the recipients don't seem to mind. Workplace gossip and gossip about social acquaintances are privately communicated. But celebrity gossip, and gossip about people in positions of power, is a multi-billion-dollar industry.

2. Reinforcing existing beliefs

Many people learn to identify themselves with a school of thought, a religion, a culture, or a set of traditions, and may not wish to be shown alternative ways of thinking or living. Naturally, such people tend to dismiss evidence of weakness in their stance. As a clear example, some people would rather hear false narratives about new rebuttals to the theory of evolution.

3. Home-team promotion and the suppression of opponents

A person who supports a sports team against another, prefers to hear positive news about their favorite team members and negative news about their opponents, even if the news is fabricated and false. The same is true of a person with a strong allegiance to a political party.

As long as there are other people who agree with us, we don't necessarily care to know the broader truth.

By design, false news tends to be more engaging and poignant than real news. Naturally, it is quoted, repeated, and discussed more often in social networks, and thus, reality has a higher potential for being overlooked.

What is referred to as "objective journalism" not only never existed in the absolute sense of the term, but the recipients of the news may not even have much desire for it. World affairs, national politics, and abstract metaphysical concepts are subjects that we tend to form our own opinions on, and as long as a journalist's article agrees with our viewpoint, we don't necessarily care about the objectivity of reporting.

Regardless of its veracity, if it is not in line with promoted policy, governments are concerned about any news that reaches their citizens. They prefer that the majority of the population receive the same political message, particularly if it has been screened or "censored" by the government. Censorship is often claimed to be for the protection of the public from "false or misleading news." Some authors such as Leo Strauss suggest that it would not be necessary for the government or any informed source to reveal the truth to the entire population but rather to a chosen elite. He referred to esoteric versus exoteric writing as writing for a select group versus writing for all, arguing that this type of selectivity has been going on for centuries[62].

Science

Science has been and continues to be our best tool for understanding the reality of the world. The well-established "scientific method" helps us accumulate new knowledge about nature, and discard arbitrary claims in favor of theories with solid foundations. Scientific progress is characterized by the continuous obsolescence of existing understandings, and their replacement by new knowledge and thinking. This is how scientific concepts evolve over time. When a theory becomes obsolete, it is usually replaced by a more precise or more broadly applicable theory. For example, "nature abhors a vacuum" is an obsolete law, but one may continue to use it in daily life instead of the more precise law that "air pressure fills a vacuum." Some people find this process difficult to accept, and instead have the false expectation that scientists and experts should be infallible,

---

[62] Hannes Kerber, Exoteric Teaching In Reorientation: Leo Strauss in the 1930s. Edited by Martin D. Yaffe and Richard S. Ruderman. New York: Palgrave, 2014.

and their opinions should never be altered. They associate the obsolescence of old ideas with a loss of credibility.

In addition to this, people have recently seen many instances of pseudoscience or "science for hire" where trained scientists deliberately deviate from the scientific method in favor of publicity or financial gain.[63] This has marred the image of science for some and given ammunition to anti-science movements that thrive in filter bubbles of the internet. This is not easily correctible by reasoning or any other means. We may be observing either an ephemeral cultural anomaly or the formation of a permanent cult of "science skeptics." Similar to religious enthusiasts, if adherents of such cults are engaged in a prolific profession and don't allow their beliefs to interfere with their work, they may still be treated as productive members of the society. In the future, Simorgh will be very tolerant of false beliefs by its human nodes as long as their quality of work is not adversely affected. On the other hand, science skeptics who want to dial back human progress will naturally not participate in the rise of Simorgh and will stay among many "unincorporated humans." This is a valid and harmless option.

> Simorgh will be very tolerant of false beliefs by its human nodes.

Reality and Fantasy

Let's explore the concept of reality in more depth and examine its foundation. We know that in the absence of an accepted basis or a set of postulates there is no *a-priori* truth. Renowned seventeenth century French philosopher René Descartes attempted to build an epistemology and a philosophy without any postulates that was to be founded on self-evident truth and logic. He started with his famous ontological expression, "I think therefore I am" (Cogito ergo sum) as the most fundamental basis of reasoning, following which all other arguments could be deduced. This statement has been thoroughly examined, and its logical nature, as well as its sufficiency as a basis

---

[63] "Manufacturing Ignorance", DW documentary, March 2021.

for reaching other conclusions, have been successfully challenged. [64,65]

It is amusing to think of Descartes' statement in the context of time: If I accept my own existence, there is still the question of how long I may have existed. The duration is undeniably finite. But is it for the duration of my memories, or as indicated on my birth certificate? Could it be that I jumped into existence a moment ago with all my memories and thoughts? Could it be that I jumped into existence in the middle of uttering the Cogito statement? We don't intend to settle these arguments here. Let us simply appreciate the fact that every reasoning and deduction requires an assumed foundation and cannot work without a set of accepted postulates.

In mathematics, which is the ultimate expression of human logic, Euclidean geometry started with a basis consisting of a well-known set of postulates, or axioms. Later, mathematicians found that violating one axiom didn't lead to discrepancies and instead led to other valid geometries known as non-Euclidean geometries. Other areas of mathematics did not have the rigid foundation of geometry, and there was a desire to restructure the entire field of mathematics and rigorously show that it was self-consistent and complete. In this context, "self-consistent" means being free of contradiction, and "complete" means being able to decide as either true or false every mathematical statement.

In early 1900s, Alfred Whitehead and Bertrand Russel tried to consolidate other mathematicians' works and build the entire field of mathematics on a firm logical basis[66]. After toiling over this for a number of years, they found it necessary to contrive controversial axioms in order to avoid certain paradoxes that stymied their work,

---

[64] Søren Kierkegaard, Philosophical Fragments; Johannes Climacus, trans. Howard Vincent Hong (Princeton, NJ: Princeton University Press, 1987).

[65] Jonas Monte, "Sum, Ergo Cogito: Nietzsche Re-Orders Descartes," Aporia 25, no. 2 (2015): pp. 13-23.

[66] Alfred Whitehead and Bertrand Russel, *Principia Mathemetica*, Cambridge Univ. Press, 1912.

and thus their initial desire of establishing a firm foundation could not be fulfilled. A number of years later in the mid-1900s, philosopher/ mathematician Kurt Gödel finally brought all such discussion to a close by proving that mathematics could not be shown to be either consistent or complete. It would take a system bigger than mathematics with which to examine mathematics. Since we don't have a superset of mathematics, we should simply avoid asking the consistency question. We have faith in mathematics because it is the most elaborate manifestation of human logic. But as we have seen, even mathematics is not immune to fundamental doubt.

Fueled by such unsettling discoveries, and by the failure of universal social philosophies such as communism and socialism to deliver the promised results of social justice[67], a philosophical movement started in late twentieth century known as post-modernism by such people as J.F. Lyotard and Richard Rorty. They questioned the validity of generally accepted theories and ideologies in Western societies and their applicability to the rest of the world. Instead, they prefer relativism over absolutism. Even knowledge is described as relative and socially influenced.

Perhaps humans have always longed for such a possibility by believing in the concept of "heaven" as a place to which they could ascend in the afterlife.

Post modernism paved the way for another marginal political philosophy in the twenty first century known as "post truth." Followers of this ideology believe that there is no objective truth; there are only narratives, and you choose to believe in the narrative that empowers you. This line of thought leads to very divisive political confrontations. Following such truth-denying ideologies is generally regarded as detrimental to society. Nevertheless, such concepts could be extended to apply to the possibility of people living in simulated

---

[67] The society is actually moving away from the utopian concept of social justice and towards greater segregation. The technological elite will break away from the rest of the society to help with the formation of Simorgh which may be a utopian outcome for a selected minority and a resource for the majority.

virtual worlds in the future. If you want to pick your own narratives, you might as well choose to live within them.

Let's explore this concept further. Since acceptance of reality requires an assumed basis, is it OK if one adopts a completely fabricated foundation? Is it OK to live in a world of fantasy? If virtual reality lets you live in a world that is self-consistent and coherent, does it matter if it bears no resemblance to the world that we currently live in? Perhaps humans have always longed for such a possibility by believing in the concept of "heaven" as a place to which they could ascend in the afterlife; a fantastic place that would fulfill all their desires.

The answer to these questions is influenced by whether the fantasy world interferes with your daily work and personal interactions. Living in a fantasy world is completely acceptable if you manage to perform all that is required of you at your workplace and in society. There is a problem if you are on a hallucinogenic drug that transports you to a dream world. The problem is that it prevents you from functioning normally in society because others cannot interact with you in a normal and predictable manner. It is different if the fantasy world is well-planned through virtual reality. Workplace tasks and your immediate human contacts in the society cannot be your own fantasy, but a variation of them presented to you as avatars within your fantasy world would be perfectly permissible and practical. The tasks that you are expected to complete in your daily job can be presented to you as challenges in an immersion game where you may even receive points or rewards for completing them.

Choosing to build partial solitude in a VR world is possible but involved. Your world must of course include your coworkers and professional acquaintances, but it may also include family members, friends, and members of the public. You can experience the entire spectrum of human emotions in your VR fantasy world, including love, envy, and hatred. Differences of opinion and interpersonal conflicts can be present. The ultimate employer, who is Simorgh, will only be concerned with each human node's completion of assigned tasks and is not expected to intervene in any interpersonal relationship outside of work time.

Another related question pertains to democratic societies, where ideally you need to have access to unbiased information before you vote for a social initiative or an individual candidate. Would it be OK if all your information came from within your own fantasy world? We may be tempted to jump to a negative answer; however it is interesting to note that this is already not far from the current state of affairs, as most people receive their political knowledge from within their own filter bubbles. Living explicitly in a fantasy world, or living in a more subtle filter bubble, have similar effects on one's ability to vote wisely. In either case, most people are still able to cast an intelligent vote on any issue related to local affairs. It is when it comes to national politics and the election of the head of the nation that the information you are being fed may have little relation to reality. This is one of many reasons why democratic societies may want to keep all voting strictly local. People are likely to be fully aware of the issues of local concern that can be decided by direct popular vote. Deciding broader national issues is best done by electing local experts who can devote more time and energy to the cause. It is only a matter of time before the fanfare of elections and voting will be replaced by routine opinion polls and behavior pattern monitoring. In the interim, as mentioned earlier, modified old concepts such as an unrestricted electoral college system to let delegates elect the head of the executive branch may need to be revisited favorably in the age of virtual reality and tailored news.

> It is only a matter of time before the fanfare of elections and voting will be replaced by routine opinion polls and behavior pattern monitoring.

Summary

The accuracy of information that we receive and the truth about the world around us are important as far as our immediate day to day activities are concerned. Everything else may be virtual, fantasy or fabricated, as long as it presents a consistent and acceptable picture to us. Human nodes of Simorgh may gradually drift into living in a pleasant virtual reality world that is part-reality and part-fantasy. Interpersonal relationships such as intimacy and conflict may also be partly mixed with fantasy, but the associated emotions are likely to remain unchanged for a long time.

Chapter 7: Automation, Robotics, and Augmented Reality

Since the early days of human civilization, people have looked for ways to take advantage of the forces of nature to reduce their own workload or to get more work done for the same effort. Taming animals for use in agriculture and transportation, as well as the invention of windmills (Figure 7-1) and water mills, are examples of harnessing natural forces to supplement human effort. This concept went through a major acceleration in the Seventeenth century when Denis Papin[68] recognized that steam from boiling water had potential for doing work. His pioneering effort, and the subsequent invention of the steam engine, led to the Industrial Revolution which reshaped human society. Trains, ships, and factories used steam to make possible what no human or animal muscle could previously achieve.

---

[68] Denis Papin (1647-1713) is known for his pioneering work on the applications of high-pressure steam.

Figure 7-1. Left: Ancient vertical windmills of Nashtifan were used 10 centuries ago to turn grinding stones. Some of them are still operable today (worldhistoria.com). Right: Modern 3-blade horizontal-axis windmill (flickr. com/photos/vax-o-matic/2621890270).

Utilization of electricity came later in human history, even though electrical phenomena were observed and documented since ancient times. Static electricity of amber, magnetism of iron, electric discharge by eels, and lightning from the clouds were known for a long time as unrelated phenomena. There was insufficient understanding of electricity to allow its generation and control for useful purposes. Ancient artifacts that look and act like batteries have been discovered (Figure 7-2); however, their utility is unknown. It is likely that they were objects of curiosity that generated, for example, a tingle in the tongue, even though practical applications such as combining multiple batteries for use in electroplating has not been ruled out.

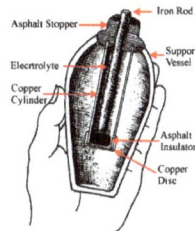

Figure 7-2. About a dozen of these artifacts known as the Baghdad Battery have been found near the ancient city of Ctesiphon in today's Iraq. They were built during the Parthian and Sassanian dynasties of Persia (240BC to 224CE). They generate about 1.2 volts of electric potential when filled with vinegar. [69]

---

[69] "Baghdad Battery also known as Parthian Battery", Elixir of Knowledge, June 4, 2014.

It was not until the nineteenth century that the electric current was harnessed and gained widespread use. Its first major application was to send messages through the telegraph. The telegraph ushered in the age of instant long-distance communication. Electric lighting and extensive use of motorized equipment followed (Figure 7-3). Soon, electricity found widespread application in all aspects of human life. In the twentieth century, one subfield of electricity named *electronics* led to the development of computers, which could be programmed to perform many repetitive calculations. Over time, the size of the computers shrank as their capabilities expanded. Now, computers can run algorithms to perform highly-sophisticated tasks without human supervision and have enabled the fast growing fields of robotics and artificial intelligence.

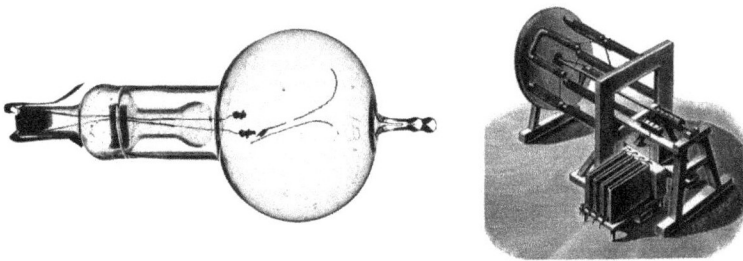

Figure 7-3. Left: One of Edison's first carbon filament light bulbs.[70] Right: One of the earliest real electric motors by Moritz Jacobi of Königsberg, May 1834.

The concept of a robot is also not new. There are numerous references in ancient literature to artificial humans and moving statues. Clockwork automatons made significant progress beginning in the Middle Ages. This included cuckoo clocks and humanoid robots such as a windup flute player shown in Figure 7-4. This flute player had the shape and the size of an ordinary man. It played a real flute by blowing air into the mouthpiece and moved its fingers to play the notes.

---

[70] Source: History of the Light Bulb, bulbs.com.

Figure 7-4. Windup flute player built in 1849 by Innocenzo Manzetti.[71]

Modern robots were first used commercially to perform industrial tasks as mechanical arms in assembly lines. One such robot named Unimate was installed in a General Motors assembly line in New Jersey in 1961 and the concept proliferated rapidly. Automobile manufacturing facilities today are heavily equipped with such robotic tools.

Service robots intended for the general public are now shaped appropriately for the service they provide; thus, they generally do not possess a human form. In 2016, a total of 29.6 million service robots were in use worldwide[72]. The types of robots that were used are shown in Table 7-1. The total number is projected to grow to 265 million units by 2026.

Type of Robot	Number	% of Total
Floor Cleaners	23,800,000	80
Unmanned Aerial Vehicles	4,000,000	14
Automated Lawn Mowers	1,600,000	5.4
Automated Guided Vehicles	100,000	0.3
Robotic Milking Machines	50,000	0.15
Humanoids, Surgical Robots, Telepresence Avatars, Exoskeletons	50,000	0.15
Total	29,600,000	100

Table 7-1. Various types of robots in use in the year 2016.

---

[71] Gianni Zacevini, "Il Primo Telefono a Milano," Divina Milano, June 3, 2021.

[72] Carlos Gonzalez, "What's the Future Role for Humanoid Robots?," MachineDesign, October 27, 2017.

Humanoid robots are currently primarily intended for use in entertainment, advertising, and as human companions. They can occasionally be seen as receptionists for publicity (Figure 7-5) or in shopping malls as guides to shoppers. Their skin texture, tone of voice, and gestures are rapidly becoming more realistic to the point where we may not be able to distinguish them as robots at first glance. Such humanoids may be perfect companions for the elderly or the disabled. They can assist with personal needs without pause and keep one company without becoming tired or bored. They can help with any task that they are capable of without any inhibition. From Simorgh's perspective, humanoid robots have limited use because they will ultimately mimic humans, and in the long run they may only be helpful to serve Simorgh's human nodes.

Figure 7-5. Humanoid robot 'Mirai Madoka' was demonstrated at the Robodex in 2017. She was developed to be a realistic looking receptionist at a Robot Development & Application Expo trade show on January 18, 2017 in Tokyo, Japan. (Source: Tomohiro Ohsumi/Getty Images)

Robot in service and under development fall into two general categories: autonomous robots and human-controlled robots. Autonomous robots carry out their responsibilities according to internal algorithms without the need for any human guidance. Examples of autonomous robots are assembly-line robots that perform manufacturing tasks, cleaning robots that sweep or mop the floors, and inspection robots that look for irregularities in systems (such as in electrical transmission lines) and fix problems as needed.

Human-controlled robots, on the other hand, are meant to augment human capabilities. They won't do much unless prompted by a human. Examples of human-controlled robots are exoskeletons, surgical robots, and avatar robots.

Eventually, people will not need to have much physical mobility at work. Virtual reality and telepresence can take care of the need to visit distant locations, and physical work can be done by automated equipment and remotely-controlled robots. For certain non-recurring tasks, however, workers will still need to be engaged in heavy physical activities, such as lifting weights and walking long distances. To help with these chores, exoskeletons are currently being developed. These robots are "worn" or "driven" by humans to augment their physical capabilities. They can also be used as personal physical therapy tools for rehabilitation (Figure 7-6).

Figure 7-6. Trexo robotics exoskeleton for disabled children.[73]

The word "avatar" is derived from the Sanskrit root *avatara*, meaning "descent," and was generally referred to the descent or incarnation of a deity on the Earth. Recently, the term avatar has been used to denote the image that a person is represented by on social media.

---

[73] Sage Lazzaro, "The 'Iron Man' exoskeleton that can let disabled children walk again", Dailymail.com, 6 September 2017.

You show your avatar instead of showing your face. An avatar robot is an extension of this concept beyond an image to a physical robot that represents its user at a remote location. Early avatar robots were simply a live video of the user displayed on a mobile monitor that could also capture video and sound. This was useful for teleconferencing and allowed the users to shift their view by remotely adjusting their avatar's camera. In the future, avatar robots will be essential to ensure that Simorgh's human nodes stay competitive with non-human nodes. An avatar robot will represent the user at a distant location, and the user will feel present and immersed in that remote environment through virtual reality.

It is virtual reality that is taking the avatar robot to the next level. Sensors and cameras on a robot provide remote visual and tactile feedback to the user. This allows the user to see the world through a headset from the perspective of the avatar (Figure 7-7). Your avatar can be sent to a remote cave, for example, and you will be able to explore that cave through the vision and senses of your avatar as if you were present in the cave.

Figure 7-7. Left: A person repairing a circuit board using a robot and virtual reality.[74] Right: controlling an avatar through a VR station (Toyota).

Virtual Reality (VR) technology received its initial boost from the gaming industry, giving players a sense of immersion into the environment of the game. This industry still remains a major

[74] Matt Simon, "Embodied Intelligence Wants to Teach Robots with Virtual Reality," Wired, November 7, 2017.

driving force behind VR. In addition to gaming, VR is now finding applications in medicine, aerospace, tourism, education, retail, and military. Virtual reality reduces the need for human mobility. As computing speeds and communication bandwidths increase, experiencing immersion in a remote environment becomes more realistic.[75] While wearing virtual reality gear, it is possible to see, touch, and interact with objects in remote locations, as if the user had been transported there. VR can similarly immerse the user in a fantasy world for entertainment or as an alternative to the "real" world. A person's view of the world and of reality may be altered with VR. As long as we can see, touch, and feel an object, it may not matter if it is purely digital in nature. Fantasy worlds may even be more inviting to some people than physical attractions of the real world. For example, one may prefer to visit Wakanda[76] rather than Machu Pichu as a tourist destination when using VR.

Computer games are playing a major role in making fantasy worlds ever more realistic. As a gamer, your avatar can interact with other players' avatars in a digital world, sometimes referred to as the "metaverse" which is a virtual shared space where your avatar can live, work, and play.

For those occasions when we don't want to be immersed in a remote or digital environment, there is augmented reality (AR). Augmented reality superimposes images or information onto what we normally see or hear (Figure 7-8). Such information may be indexed to a location, it may be context-driven, or it may be independent of what we observe. Location-indexed AR is, for example, a gaming application in which one must find virtual objects hidden in

> When we can see, touch, and feel an object, it may not matter if the object is purely digital in nature.

---

[75] While wearing a VR headset, the system needs to respond quickly to any movement that you make. The image presented to you in the headset needs to change when you move your head. This quick, unpredictable rendering of images and other signals requires very high speed computation and communication. Any lag in the response or lack of detail in the surroundings can be disturbing to the user.

[76] A fictional hidden but highly advanced African nation as depicted in the Marvel Franchise film *Black Panther*, 2018.

the user's immediate vicinity[77]. Outside of gaming, location-indexed AR allows retail stores to advertise as you walk in front of them, or it can let users see new possibilities and outcomes, such as the consequences of a development project before it is implemented. One can see a building in its actual location before it is built. This allows the design to be reviewed and modified for a better fit to the neighborhood.

Context-driven AR may superimpose names and background information near the face of each person you meet at a gathering, or it could superimpose measurements of objects you see, or fact check statements that you hear. Context-driven AR helps to make meetings more productive, accelerate decision-making, and facilitate tasks by providing contextual information.

Finally, independent augmentation may be the continuous displaying of time, temperature, or stock market index in the corner of your field of view. Such AR technologies are currently implemented in "heads up displays" in cars, where the information is projected on the windshield. It may also be projected on eye glasses that you wear, or it may be directly projected into your eyes through a technology known as "retinal projection" or "Virtual Retinal Display" (VRD).

Figure 7-8. Left: Augmented reality game known as Pokémon Go. Virtual "pocket monsters" can be found in the user's vicinity.[78] Right: Animation projected on the user's desktop.[79]

---

[77] Pokémon Go is one such game.

[78] Gianluca Busato, "Why You Should Start Using Augmented Reality (Ar) and Gamification?," Medium, March 27, 2018.

[79] Nitin Agarwal, "The Idea of Augmented Reality in App Innovation," Wildnet Technologies, April 30, 2019.

Retinal projection is a method of adding information to your field of view by using the retina of your eye as the display screen. One approach, shown in Figure 7-9, uses a transparent holographic reflector on your glasses. The reflector is not visible, but it redirects light from small lasers in the frame of the glasses into your eye, forming an image directly on your retina. Such images are much clearer and sharper than any image you might see on a high-quality television screen. The reason is that the TV screen still needs to be imaged onto your retina through your eye's cornea and lens that may add aberrations to the image.

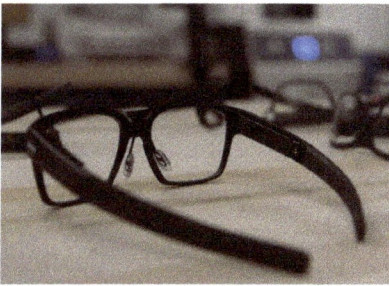

Figure 7-9. Intel's Vaunt AR glasses inconspicuously project information directly into your right retina.[80]

AR eyeglasses will soon inherit the capabilities of smartphones. Currently, most people's connection to the internet is inconveniently through the smart phones that they carry. It is apparent that in the near future the preferred connection method will be through augmented reality eyeglasses. This way, connection to the internet can be uninterrupted all day, yet easily stopped by removing the eyeglasses if preferred by the user.

Our ears and eyes are currently our main interface ports between the human nervous system and the internet. This means that all electronic information first needs to be converted to either images or sound before being received by a human. This inherent inefficiency has sparked an interest in bypassing this media transition and making a direct electronic connection to the human nervous system. Direct neural interfaces (DNI) or brain-computer interfaces (BCI) may become a reality in the near future, as several companies are

---

[80] Sean Hollister, "At Last -- Smart Glasses That Don't Look like Borg Headgear," CNET, February 5, 2018.

currently conducting research on the subject.[81]

One complication for BCI is that the brain's interpretation of any neural signal depends on which part of the brain it goes to. Artificially routing an external signal to the correct part of the brain is theoretically possible but it is not yet practical. As a result, establishing a "data port" to the brain is not as simple as electronically accessing a few neurons. Significantly more insight into the working mechanisms of the brain is needed before any practical BCI can be developed. However, the work to address this challenge is ongoing, and already some rudimentary movements of robotic arms can be controlled by human thought.

Eventually, brains and computers will be able to directly communicate, allowing exoskeletons and VR gear to merge with their users. Gene editing will help the merging of electronics with humans by allowing the nervous system to be more adaptable to external transmitters and receivers. Gene editing will not only augment human capabilities but may also alter human appearance according to the profession and specialization that they need to adapt to. Most human nodes at work in Simorgh's body will be genetically-modified humanoids (GM hybrids) wrapped in mechatronic contraptions (see Chapter 13).

Summary

Of the natural phenomena harnessed by humans, electricity has proven to be the most versatile. It lights our nights, heats and cools our homes, and powers our appliances. But above all, it is used in logic circuits that perform calculations, data processing, and automated motion. The latter has led to autonomous and human-controlled robots, avatars, and exoskeletons. Virtual reality and augmented reality, enabled by advanced data processing, are essential for people to compete with emerging intelligent electronics. Future human nodes of Simorgh will be highly-specialized, genetically-enhanced humanoids with permanent connections to electronic gadgets and Simorgh's nervous system that is the future of the internet.

---

[81] Neuralink, BrainRobotics, Kernel, Brain Co, and Mindmaze, to name a few.

Chapter 8: Augmented Humans: Bionics, Artificial Intelligence, and Gene Editing

Ancient legends of heroes with superhuman abilities accentuate our persistent craving for overcoming weaknesses and undesirable traits of humans. The body of Achilles, except his heel, was rendered invulnerable by being dipped in the river Styx as an infant[82]. Samson's source of his superhuman powers was his hair, according to the Old Testament. Alchemists in search of the "Elixir of Life" took seriously the quests for extended life and immortality. In Japanese mythology, the character Yao Bikuni, or the "800 nun," was a girl who accidentally ate Ningyo[83] meat and was "cursed" by immortality. She was finally relieved of the burden of life at the age of eight hundred. In many of these stories, the person with superhuman abilities becomes unhappy, is eventually defeated, or develops undesirable characteristics, preaching to the readers to be content with "normal" human weaknesses.

Legends typically depict a sudden acquisition of extraordinary capabilities by magic. In reality, there has been a more gradual and

---

[82] In a similar story, as told in *the Epic of Kings* collected by Ferdowsi, the invulnerability of Esfandiar does not extend to his eyes, which were closed during his immersion into the divine water as an infant.

[83] Ningyo, or "human fish," is a mermaid-type character in Japanese mythology.

sustained source of ever-increasing abilities for humans through the development of tools. Early invention of the wheel and the control of fire created tremendous advantages for people and set us on the path of technological advancement at ever-accelerating rates. Aside from tools for general use, there are many gadgets that are tailored to a specific user for a specific purpose. Popular examples include widely used vision-correcting eyeglasses and hearing aids. Prostheses for the injured and disabled have also been in use for centuries, albeit in a rudimentary form until recently. Now, prosthetic limbs with fine motion control, a realistic look, and even stimulation by the nervous system are all within reach (Figure 8-1).

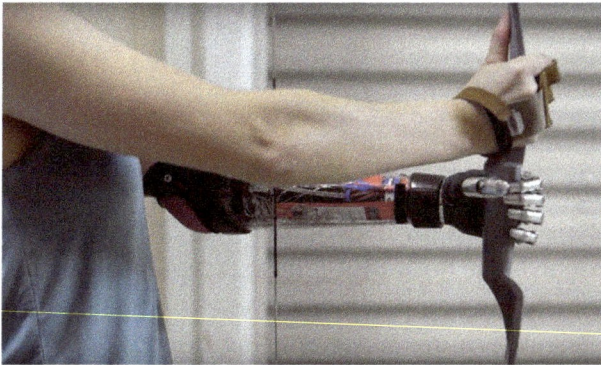

Figure 8-1. Prosthetic limbs are beginning to be equipped with bilateral connections to the nervous system (designboom.com).

A phenomenon known as myoelectricity has been utilized in prosthetics to enable a degree of mental control. This phenomenon relates to the fact that live remnants of muscles lost due to an injury can still respond to mental attempts at moving a missing limb. Sensors can detect the electrical signals on the tissue and stimulate a corresponding action in the prosthetic arm. With practice, the user can learn to make the desired movements, although without any feeling of touch. Sensory feedback, or the sensation of touch, is also slowly evolving. The field of neuroprosthetics is actively pursuing the development of bilateral connections to the nervous system. Prototypes of prosthetic arms have been developed that are not only controlled by the user's brain, but also provide a rudimentary tactile feedback.[84]

Robotic exoskeletons are powered suits that enhance the physical capabilities of their users. The operator can lift heavier weights, walk longer distances, and in general have more physical power and stamina. This is a realistic technological path to transforming a person into what could be called a "superhuman." This technology will soon be within public reach (Figure 8-2). Armed forces in many countries have shown interest in the development of powered exoskeletons that could potentially transform anyone into a formidable super-soldier while the battery lasts!

Wide availability and versatility of robotic exoskeletons or "power suits" will accelerate as the storage capacity of their portable power sources increases. Advances in battery pack and fuel cell technology are gradually removing such power storage constraints.

Figure 8-2. Robotic exoskeletons add strength and protect the human body. Left: the user is lifting two bowling balls with little effort. Right: walking assistance is helpful to people who carry weight in their daily work.[85]

Storage densities of batteries have doubled in the past ten years, as shown in Figure (8-3), and their average price has dropped by almost a factor of five. Storage density is expected to increase by another factor of two over the next four years, putting a storage capacity of

[84] Allen, Daniel. "The Rapidly Developing Field of Neuroprosthetics." MedicalExpo e-Magazine, July 7, 2021.

[85] "Robotic Exoskeleton, the External Skeleton Made for Humans," Ozwana, accessed July 27, 2021.

1kW-hour per liter of battery volume within reach. With that much storage capacity, an exoskeleton carrying a six-liter battery pack can boost its user's physical capabilities by the equivalent of one horsepower over an eight-hour period of continuous use.[86]

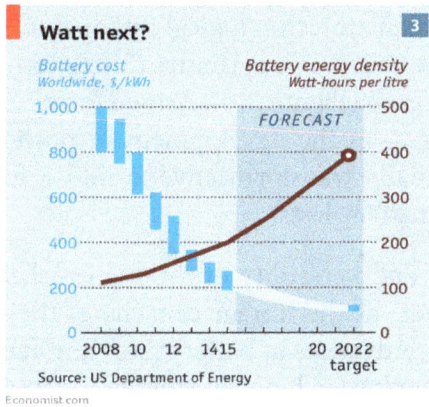

Watt next?

Battery cost
Worldwide, $/kWh

Battery energy density
Watt-hours per litre

FORECAST

2008 10 12 14 15   20 2022
target

Source: US Department of Energy

Economist.com

Figure 8-3. Capacity increase and price drop of batteries within the last ten years.[87]

A person wearing an exoskeleton might be considered a superhuman, or he might be a disabled person who can function normally with the help of the power suit. In either case, the human-exoskeleton combination will evolve into a "human-robot hybrid" and will provide humans with abilities they would not otherwise have.

Hybridization with robots is not the only way to enhance human capabilities. Another emerging route is the modification of the human genome. A gene-editing technology generally referred to as CRISPR/ Cas9[88] has brought this prospect close to reality. Editing a DNA molecule is now possible by targeting an edit location, performing a

---

[86] Fuel cell technologies are also evolving alongside batteries and have the capability to store sufficient energy in a small volume with non-polluting fuels. They will likely become affordable in the near future. Currently, their use is on the rise, particularly to power forklifts and similar equipment for warehouse material handling.

[87] "After electric cars, what more will it take for batteries to change the face of energy?", The Economist, 12 August 2017.

[88] CRISPR is an acronym for Clustered Regularly Interspaced Short Palindromic Repeats that were initially observed in acquired immune systems of bacteria.

double strand cut of the DNA using the enzyme Cas9, and inserting another sequence in its place (Figure 8-4). The procedure is low-cost and on its way to becoming as simple as word processing.

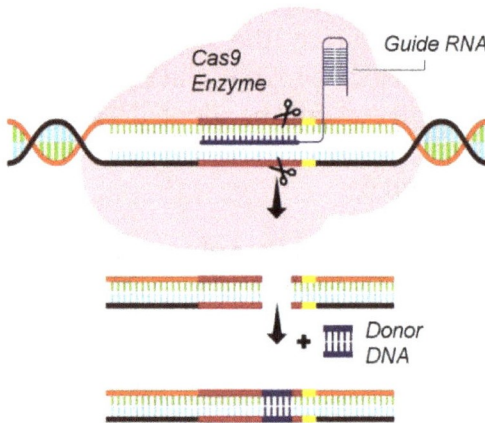

Figure 8-4. CRISPR/Cas9 gene editing sequence: The guide RNA identifies the editing location. The Cas9 enzyme performs a double strand break. The DNA is repaired with the inclusion of the donor sequence.

Human Polymorphism

Modification of the human genome will initially be driven by the desire to suppress genetic diseases such as diabetes and autism, but it is only a matter of time before we start to augment various characteristics of healthy humans. It may start with minor improvements in physical or mental capabilities, but eventually, different genetically specialized people will emerge. It is important to note that human genetic augmentation cannot be an arbitrary selection from a menu. Most modifications will have unintended consequences, so you cannot select the traits of your offspring at will. Overtime, compatible sets of enhanced traits will be identified, and only a few types of augmented humans will emerge. This will be similar to current sex differentiation in people, or "caste polymorphisms" in insects with stable and compatible characteristics.

Existing laws prohibit genetic alteration of humans in many countries, and for a good reason. A lot of ethical and procedural issues need to be resolved before such experiments can proceed safely. This prohibition will inevitably be lifted and replaced with procedural guidelines for alteration. Unreasonably restrictive and conservative laws may delay the start of genetic engineering of humans, but technological progress will eventually prevail, and new legal frameworks will certainly be adopted.[89]

Artificial Intelligence

Artificial intelligence (AI) is a field of computer science with the aim of building machines that can function without relying on detailed instructions from humans. Early successes in AI came in the form of "Expert Systems" that follow a human expert's method of thinking toward solving a problem. For example, to diagnose and repair malfunctioning equipment, an expert human may ask questions with binary answers in order to narrow down the location of the fault. A computer algorithm written as an expert system imitates the expert human by following the same procedures and methods, but it is incapable of developing its own methodology. Such expert-systems allow the skill set of one expert or a group of experts to be copied and used at multiple locations.

The field of AI took a major step beyond expert systems when "machine learning" techniques were developed. Machine learning can generate its own abstract rules from a large number of inputs and apply them to more general situations. For example, if you want to teach a machine how to recognize a cat, you can take an expert-systems approach and have the machine go through a set of questions and criteria to qualify an image as that of a cat. On the other hand, with machine learning you show the machine many different images of cats and allow the machine to find its own recognition method. In many cases, the method that the machine adopts has no similarity

---

[89] In general, technology regulation may be used to either inhibit or stimulate progress, but it can rarely be used to stop the progress in areas where large capital investment is not required.

to a human's recognition approach. Machine learning is capable of modifying its own procedures in order to reach the desired goal.

A more recent subset of AI known as "deep learning" has multiple layers of machine learning algorithms (Figure 8-5). Each layer processes a different aspect or interpretation of the input data. Such algorithms are also called artificial neural networks. Deep learning is developing the capability of working with unstructured input data, whereas machine learning requires structured or labeled data for training.

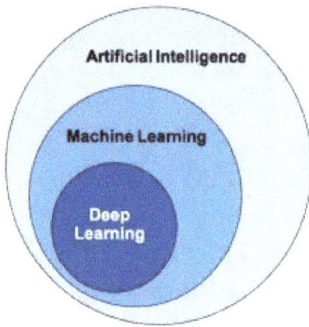

Figure 8-5. Machine learning and deep learning are subfields of Artificial Intelligence (AI).

To date, AI successes – at least those that have garnered the most attention – have centered on outperforming human champions in a variety of games such as chess, Go, and Jeopardy, or delivering satisfactory performances in areas such as driverless vehicles. What these cases have in common is their limited scope. They are each focused on performing one limited task. In contrast, in our daily lives, we all face a variety of chores and challenges that, while individually simple, are collectively beyond the capabilities of today's AI technology. The branch of AI that is attempting to similarly broaden the scope of intelligent machines is known as artificial general intelligence, or AGI. Such AGI machines receive diverse inputs, such as all environmental stimuli perceived by humans, and attempt to address each one appropriately. There is a milestone set in the future when AGI technology matches the level of human intelligence. This milestone is referred to by some experts as the "singularity" in AI. What will develop after the singularity is "Super Intelligence," when AGI machines start to program their own next generations with limited or no input from humans. The concern is that from that point on, machine intelligence will advance at such a rapid pace that humans will be left in the dust. Soon thereafter, its development direction may become totally incomprehensible by any human.

There is the philosophical question of whether AI will mimic human intelligence. Currently, there are distinct differences between the two. AI does not recognize causality; it is incapable of generalization and abstraction; and until recently, it was not capable of unsupervised learning. Over time it will acquire some of these capabilities, but it doesn't need to mimic human intelligence to be effective and capable of making rapid progress.

> The threat of AI and autonomous robots to humanity will be real only if we fail to plan and adapt.

There is a widely used AI training method known as reinforcement learning, in which rewards and penalties are assigned to approaching and solving various problems. The programmer designs the reward policy, but it is up to the computer to find the approach that will maximize the reward. We commonly use reinforcement learning in our daily lives and in children's education. It is instructive to note that all living organisms also learned their life skills from a simple reinforcement learning method sometimes referred to as the survival of the fittest. This is the only programming that all life received from nature. The problem to solve is how to ensure the survival of one's species in view of competition and limited resources. The reward is survival, and the penalty is extinction. Simple life forms developed simple survival skills, while more complex life forms developed more sophisticated tools for survival; tools such as consciousness, pleasure and pain. Today's AI receives much more detailed and specific reward policies. It is impossible to predict what skills the computers will acquire if sometime in the future, their survival becomes their dominant reward policy, as it is in nature.

AI and autonomous robots are two technologies that some see as threats to humanity. The threat will be real only if we don't plan ahead and fail to adapt accordingly. Table 8-1 lists three such concerns with a suggested response to each that will ensure continued human involvement and superiority. We need to ensure that human involvement always enhances technology and helps it remain superior to technology without humans.

What skills would future computers acquire if survival becomes their dominant reward policy?

When outperformed by a new form of automation, it is common for people to initially feel threatened before learning to embrace it, or even become addicted to it. Computers are already performing many tasks much better than any human. That's why, for example, we let our navigation system decide our driving routes. We see our navigation system as an extra capability that we have. We take ownership of it and we merge with it. The technology changes us and gives us more capabilities. This is the proper way to react to modern technological advances in general.

Trend	Human-Centric Alternative
Autonomous robots	Humans with robotic exoskeletons
Computers with Artificial Intelligence	Humans with Brain-Computer-Interface
AI superiority to human intelligence	GM Human-Computer Hybrids

Table 8-1. Proper human response to the listed technology advances: Autonomous robots may render some human work superfluous, but humans with robotic exoskeletons will stay competitive. Similarly, BCI and genetic modification will ensure that when merged with technology, humans will retain their superiority over technology alone.

The last entry of Table 8-1 refers to genetic engineering, which will allow humans to keep up with the fast pace of development in artificial intelligence and machine learning. Genetically enhanced humans may develop super brains with vastly higher analytical capabilities. Bilateral electronic connections to the human nervous system will develop, and as a result, humans will merge with computers and robots and will evolve together in Simorgh's body. A large population of humans will be left out of this development path either by choice or by chance: voluntarily or by social exclusion. They may eventually be deemed as an underdeveloped species or as "primitive societies." However, they

Most people will be left out of this development path. Their relationship with Simorgh will be similar to the current relationship of bacteria with the human body.

can still benefit from Simorgh either parasitically or synergistically. This is similar to the current relationship of good and bad bacteria with the human body (Figure 8-6, and Chapter 13). Needless to say, Simorgh will have to maintain some control over primitive societies in order to prevent harm and maintain safety in its own vicinity.

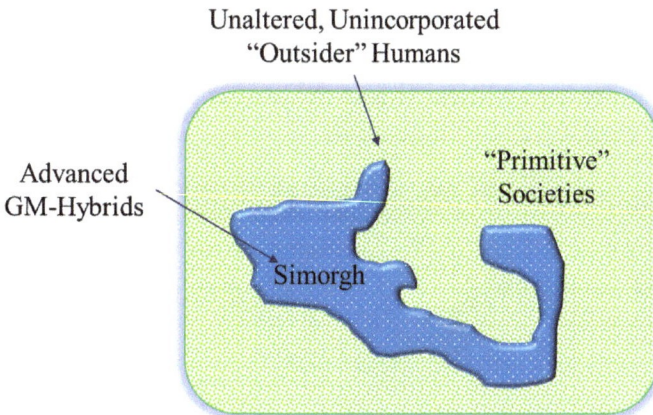

Figure 8-6. Simorgh's body cells (nodes) will include many Advanced Human-Robot Hybrids (GM-Hybrids) who are genetically enhanced humans augmented by computers and robotic technologies. Most other humans will live on the outside of Simorgh's network. They will have limited robotic help and an unaltered genome. They can benefit from Simorgh's presence just as bacteria benefit from the human body.

Summary

Rapid progress in artificial intelligence should not be alarming to us. Advances in augmentation technologies and genetic engineering will allow humans to progress just as rapidly. A specialized minority of the human population is about to split away from the rest and become body nodes of the emerging Simorgh. This minority (GM Hybrids) will undergo genetic modification and will seamlessly merge with technology. The rest of the population (Unincorporated Humans) will continue to live with a lesser focus on technology, forming what we may label as "primitive societies." These societies can coexist with Simorgh and benefit from its presence. Their role in relationship to Simorgh is analogous to that of bacteria in relationship to the human body.

Chapter 9: Large-Scale Regulating Systems

The sun's total average radiative power is about 3.86×10^{26} watts (386 followed by twenty-four zeros)[90] – a massive number that doesn't fit in our imagination. Only about half a billionth of that power reaches the Earth, which is approximately 1.74×10^{17} watts. Except for geothermal, tidal wave, and nuclear energies, all other energy on Earth originates from the sun. Some is converted to wind energy, while some is stored into plants, biomass, fossil fuels, and so on. The total annual energy consumption by humans (6×10^{20} joules or 600 EJ in 2019) is equivalent to the energy supplied to the Earth by the sun in fifty-seven minutes. Our energy consumption has grown by a factor of four in the last fifty years (Figure 9-1). If this rate continues, which it may not,[91] in a little over three centuries humans will be consuming the equivalent of the entire solar power impinging on Earth. Of course, we can't block all of the sunlight and intercept all of the solar radiation arriving to Earth, without severe consequences. It is possible to intercept a significant amount of additional solar energy with outer space mirrors without casting a shadow on Earth. This extra sunlight can be redirected to the Earth or consumed by outer

---

[90] "The Solar Constant," Australian Government - Bureau of Meteorology, Space Weather Services, accessed July 27, 2021.

[91] The trend will most likely flatten. See Appendix D.

space-based industries. It is also safe to assume that nuclear energy, both by existing technology of fission and future technology of fusion, will provide a major portion of human society's energy needs. Energy consumption at this scale cannot be without consequences, particularly on the planet's climate.

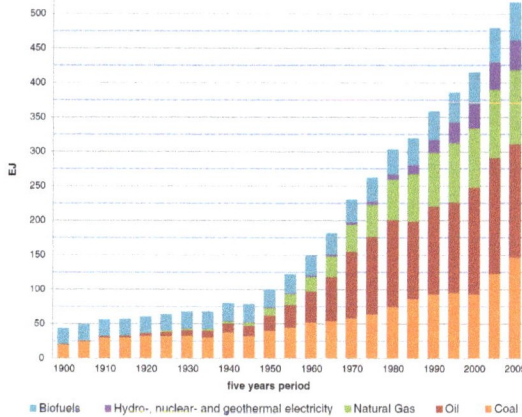

Figure 9-1. Worldwide energy consumption.[92]

It is important to note a basic physics fact that consumed energy never disappears. Instead, it is converted from one form to another, and any part of it that is not stored is eventually converted to heat. So, the more energy we consume, the more

> Consumed energy never disappears. Instead, it is converted from one form to another.

heat we deliver to the environment. In order to explore the outcome, let's have a simple thought experiment. Any mechanically isolated warm object that does not receive energy from the environment, loses heat by radiation. This loss of heat cools down the object. At low temperatures, the amount of radiation is not high, and the emission is in the invisible infrared wavelength range. As the temperature rises, the amount of radiation increases, and its average wavelength shortens until it eventually becomes visible and may be seen as a glow. A stable temperature is maintained when the heat[93]

---

[92] Kostas Bithas and Panos Kalimeris, "A Brief History of Energy Use in Human Societies."

[93] Power is defined and the rate of energy used or delivered.

being carried away by radiation is compensated by an equal external heat flux directed into the object (Figure 9-2). If more energy flux is added to the object, its temperature rises, causing it to radiate more until the additional radiation rate equals the additional input power, and a new higher equilibrium temperature is established.[94] If the cooling radiation is somehow blocked, then the temperature can rise to much higher levels. An object that is surrounded by an insulating blanket retains more of the heat and reaches higher temperatures.

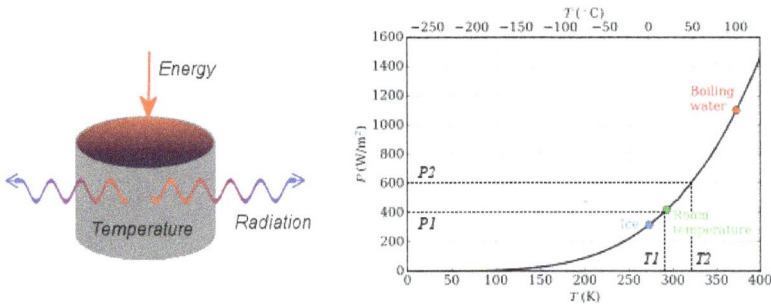

Figure 9-2. An isolated object has a stable temperature when the energy it receives is equal to the energy it radiates.[95] When the rate of energy delivered to the object goes to a higher level, temperature and radiation increase and a new stable higher temperature equilibrium is established.

A similar analysis applies when the object is planet Earth. As more heat flux is added to the Earth by human energy consumption, the atmosphere warms and a slightly higher equilibrium temperature is reached while the energy consumption lasts. However, with extra carbon dioxide from fuel burning, and extra water vapor from evaporation, the atmosphere will absorb more of the cooling radiation that would normally escape into space. The extra carbon dioxide and water vapor act as a blanket. This is known as the greenhouse effect and can amplify the temperature rise.

Let's look at this from another perspective. Currently, about thirty percent of the sunshine impinging on Earth is reflected back into space (Earth's albedo is ~ 0.3). The remaining seventy percent keeps

---

[94] For simplicity, it is assumed that the object is isolated with no contact with other objects, and that there is no phase change within the object, e.g. freezing or melting.

[95] The amount of the radiation that cools the object increases with the fourth power of temperature (Stephn-Boltzman law).

the soil and waters warm and runs solar powered processes such as photosynthesis. The Earth re-radiates the equivalent of all the non-stored energy back into space as infrared radiation, so the net temperature rise is zero (Figure 9-3). If the temperature of the Earth goes up, for example by the cyclic increase of solar

> The real danger is this runaway climate change that may render the entire planet uninhabitable by humans, not a few degrees of warming that may only change the map of climate favorability on Earth.

radiation, the temperature rise will be moderate. The reason is that the Earth's infrared radiation into space will increase to stabilize the temperature. If instead of adding more heat, or in addition to it, a blanket of greenhouse gasses forms around the Earth and prevents the cooling radiation from escaping into space, the temperature rise will not be moderate, since more heat tends to generate more greenhouse gasses (such as water vapor and methane from melting permafrost), and may lead to a runaway effect such as what has happened on the planet Venus. The real danger is not a few degrees of warming that may only change the map of climate favorability on Earth, but the potential runaway climate change that may render the entire planet uninhabitable by humans. Unfortunately, the tipping point or the temperature above which the runaway transition occurs is unknown.[96]

In the future, some warming of the atmosphere may be inevitable. More direct and widespread use of solar power will reduce the Earth's reflectance, or albedo, and will increase the atmospheric temperature. The use of stored energies such as geothermal and nuclear will also add heat to the overall balance. Any such warming without a greenhouse blanket will be confined to a few degrees. This will cause polar ice caps to melt and ocean levels to rise. Consequently, some coastal regions will be flooded, and some currently frozen lands will

---

[96] The effect of increased cloud formation on the overall balance of heat has not been fully simulated. Clouds reflect a lot of sunlight back into space and prevent heating by sunlight. At the same time, they blanket the Earth and keep it warm. If the net effect is toward cooling the Earth, they will have a moderating effect. Otherwise, they will accelerate the warming trend.

become desirable for living. On balance, the total inhabitable area of the Earth will increase. The average temperature may stabilize at a somewhat higher level, and some people may see this as favorable. However, the story may not end here. Much caution and very careful planning are needed because the temperature rise itself will trigger the releasing of greenhouse gasses, and the melting of polar ice caps will reduce the Earth's albedo. This combination has the potential to trigger the dreaded runaway heating.

Systematic management and control of temperature rise may be done by large climate engineering projects such as carbon sequestration, seeding the clouds to induce more rainfall, and limiting the energy received from the sun by reflecting sunlight back into space. However, these approaches take a long time to master, since each will offer a partial solution and will have unintended consequences. The immediate tools at our disposal are to pursue higher energy efficiency and population control to flatten the curve of human energy consumption, together with environmentally friendly technologies to curtail the generation of greenhouse gases.

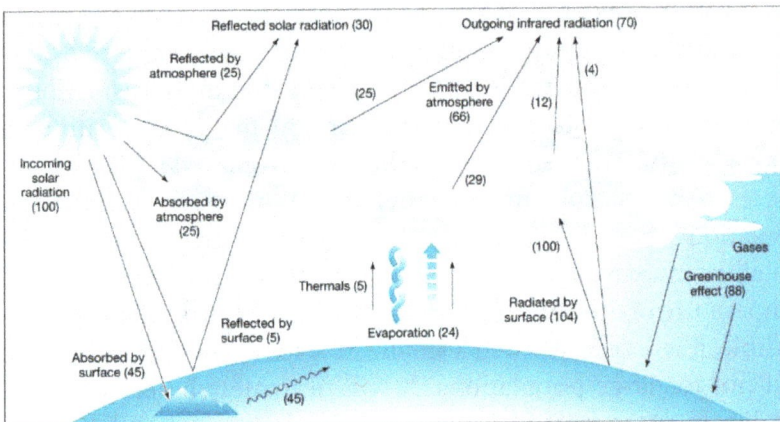

Figure 9-3. Atmospheric energy balance. [97]

Would humans or Simorgh survive if global warming is not controlled? Simorgh's non-human nodes may be able to adapt to

---

[97] Camilo Rada, "What Is the Difference between Radiation Balance and the Global Energy Balance?," Earth Science Stack Exchange, March 11, 2019.

somewhat higher temperatures, while humans have a narrower comfort zone. Electronic circuits generate heat and often need to be cooled for proper performance and good reliability. Most common electronic devices need to operate in the ambient temperature range of zero to eighty-five degrees Celsius (32° to 185°F). Specialized high-temperature circuits can operate at temperatures of up to a few hundred degrees Celsius. The human body, in contrast, regulates its own temperature around thirty-seven degrees Celsius, and has the narrow survival body temperature range of 34° to 44°C. If after the emergence of Simorgh the Earth's temperature becomes too high for human survival, it will tip the scale against the employment of human nodes by Simorgh in favor of electronic nodes. Simorgh may continue to thrive, but the fate of humans will be more precarious.

Can we build enclosed, air-conditioned quarters, communities, or even cities for the protection of humans? The following facts must be kept in mind in order to grasp the ramifications of this: (1) Coldness is the absence of heat, and unlike heat, it cannot be generated. You can only pump heat out of a closed volume and dispose of it elsewhere along with the extra heat that your effort generates. For example, a refrigerator heats up the room by more than the amount of heat that it removes from the inside. (2) For a given amount of insulation and power consumption, there is a maximum temperature difference that can be maintained between the target volume and the ambient. Therefore, as the ambient gets hotter, it takes more energy and heavier insulation to maintain a constant pleasant temperature inside. Installing massive air conditioning plants will further increase the temperature of the atmosphere. So even if building such protection is feasible, it will only work in a small space for a small group of people and not for large populations. It will ultimately be easier to allow the Earth's extra heat to escape into outer space by radiation than maintain areas of air conditioned living space on Earth. Allowing heat to escape requires removing the blanket of greenhouse gases. If a single Simorgh emerges on Earth, maintaining a comfortable atmospheric temperature will be one of its high priority goals. If multiple, non-cooperating Simorghs emerge, the outcome will be more uncertain.

Climate change is not the only human-made problem facing the planet. Other issues such as potable water scarcity and waste disposal are changing from local problems to matters of global concern. As we gain more control over the planet, we must treat it more like our own house, and need to provide for its heating, cooling, water, power, and waste management. This calls for global preservation, geo-engineering programs, and the development of large-scale regulating systems to be eventually operated as body organs of Simorgh.

Water Management

Population growth and the ensuing demand for fresh water has caused many rivers and lakes around the world to go dry. Noteworthy among the drying lakes are: Aral Sea in central Asia that used to be the world's fourth largest lake; Lake Chad in west central Africa; and Lake Urmia in northwest Iran. There is no natural self-correction to this trend, and effective management of major water resources is essential to human survival. In order for Simorgh to care for its human population without reservations, effective solutions for water resource management need to be readily available prior to Simorgh's emergence. Most freshwater consumption by humans is for agriculture, as seen in Figure 9-4. With alternative food production methods in the future, the strain on freshwater resources may be relieved. Currently and in the short term, major methods available for water management are: (1) water-efficient irrigation and farming; (2) water recycling; (3) sea water desalination; (4) deep aquifers and fossil water; (5) large scale redistribution networks; and finally (6) local climate control to enhance water supply. Each method's cost-benefit tradeoff is geography-dependent, and it is likely that a combination of solutions would be optimal for any location. Let's examine each one of the listed tools separately.

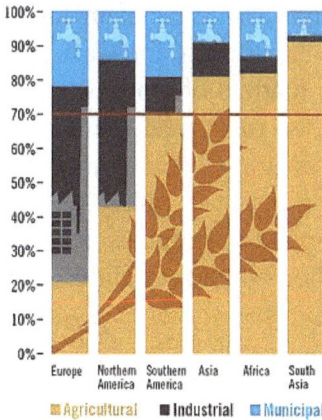

Figure 9-4. Water consumption by various sectors in different regions of the world. (Source: Globalagriculture.org)

1. Water-Efficient Irrigation and Farming

Traditional irrigation methods, such as furrow irrigation, are wasteful in the sense that most of the water used is not consumed by plants. Efficient irrigation methods conserve water and increase the productivity of farms per quantity of water used. Drip irrigation and its variations deliver measured amounts of water to individual plants and prevent indiscriminate soaking of the soil where there are no plants to absorb the water. Precise irrigation scheduling based on weather conditions prevents overwatering and underwatering of plants. Indoor and hydroponic farming (Figure 9-5) further decrease water use by preventing water loss to the ground and controlling the water evaporated from plants. Typically, drip irrigation consumes about one quarter of the water used in traditional furrow type irrigation, while hydroponic farming uses one twentieth of that water[98]. Furthermore, the climate control afforded by indoor farming allows traditionally seasonal crops to be available year-round.

> Drip irrigation consumes about a quarter of the water used in traditional furrow type irrigation, while hydroponic farming uses one twentieth of that water.

---

[98] [G. Pollard J. Ward and P. Roetman "Water Use Efficiency in Urban Food Gardens: Insights from a Systematic Review and Case Study", Horticulturae 4(3) 27].

Figure 9-5. Tomatoes grow year-round at Mastronardi's greenhouse in Coldwater, Michigan. (Source: Rayan farms, "Hydroponics cherry tomato.")

2. Water Recycling

Currently, fresh potable water that is suitable for human consumption is also used in households and farms for irrigation. This is a wasteful practice given our dwindling freshwater supply and will inevitably need to be rectified. Reclaimed or greywater[99] may be used for irrigation and any household application that does not involve direct human contact. Reclaimed water use may be implemented within a household independent of others, or cities may employ reclaimed water distribution systems (Figure 9-6). Of course, supplying two or more water lines to each household, as well as the corresponding sewage lines to reclaim used water, is expensive to implement in existing communities and may only be feasible in new residential developments. In existing communities, it is more economical to pipe the reclaimed water to large institutions or farms for irrigation. Using reclaimed water for irrigation has advantages, such as providing needed nutrients to plants. But it is not free of complications either. For example, prolonged use of greywater may increase the salinity of the soil. Limited filtering and desalination of greywater may be needed.

---

[99] Greywater is used water from homes or businesses that does not contain human waste and is diverted and sometimes treated for reuse in landscapes and other non-human contact applications.

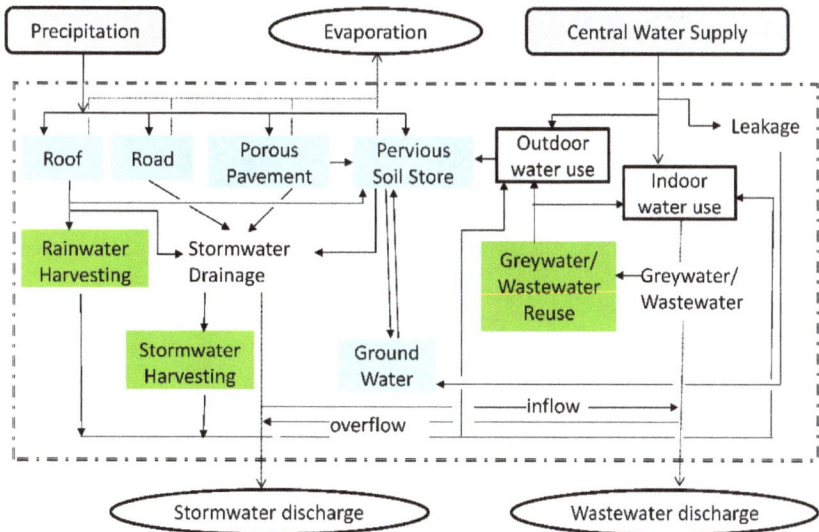

Figure 9-6. A proposed urban water distribution and recycling network (Australia).[100]

3. Sea Water Desalination

The idea of desalination and the desire for it are not new. The philosopher Aristotle explained how water distillation removes its salinity: "Saltwater, when it turns into vapor, becomes sweet, and the vapor does not form into saltwater again when it condenses." A sixth century commentator of Aristotle named Olympiodoros reports: "Seamen, when they lack fresh water at sea, obtain it as follows: they boil the sea water, then place large sponges over the boiling water to collect the rising vapor. When they squeeze the sponges, the water obtained is fresh."[101] Despite the interest, a commercial desalination plant was not built until 1881, when a seawater thermal distillation plant was constructed in Malta to desalinate sea water for the local population. Currently, large capacity desalination plants exist around the world to produce potable water for human consumption.

---

[100] Mukta Sapkota et al., "An Overview of Hybrid Water Supply Systems in the Context of Urban Water Management: Challenges and Opportunities," Water 7, no. 12 (December 29, 2014): pp. 153-174.

[101] Giorgio Nebbia and Gabriella Nebbia Menozzi, *A Short History of Water Desalination* (Milan: Azienda Grafica Italiana, 1966).

Most methods of desalination fall within the following categories: thermal, electrical, and physical. In thermal distillation – which is the oldest method – water is boiled or evaporated, and the collected steam is cooled to liquify. Some systems use low-pressure vessels to reduce the evaporation temperature, thereby reducing the amount of energy used in the evaporation process.

Another method of desalination utilizes electricity to separate the water and salt. Dissolved salt ions that are electrically charged are moved through a permeable membrane. This method is also known as electrodialysis (Figure 9-7). The electrical energy used in the process depends on the concentration of salt in the water. Therefore, it is more economical to partially-filter the sea water ahead of time.

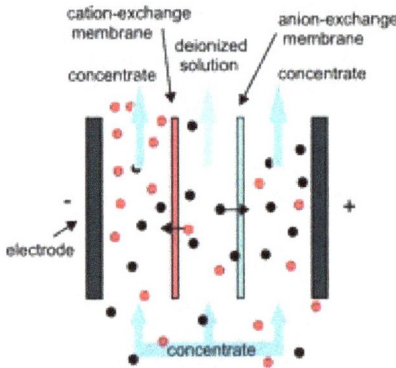

Figure 9-7. Electrical desalination, or electrodialysis, employs membranes that only allow ions to pass through. What is left behind is purified water.[102]

Finally, there are physical methods that drive water through a membrane to filter out the salt. These methods are also known as reverse osmosis (RO) technologies. Salty water is pushed through the RO membranes at high pressures of about 1000 pounds per square inch (seven mega pascals). Membranes and meshes of different types are designed to form cylindrical layers, as shown in Figure 9-8. The purified water seeps into the center perforated pipe, while the salty concentrated water moves out parallel with the layers. Some of the energy used to pump the water at high pressure through the membranes may be recovered from the pressure of the outgoing water. Since more energy is used for higher salt concentration,

---

[102] S. H. Roelofs, et. al., "Microfluidic desalination techniques and their potential applications," Royal Society of Chemistry, Issue 17, 2015.

it is more efficient for physical methods to use multiple stages of purification with successively finer grades of membranes at each stage (Figure 9-9).

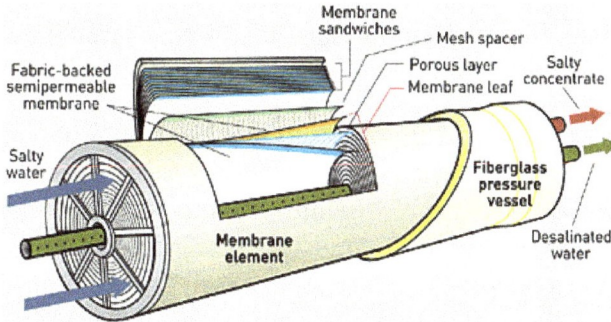

Figure 9-8. An example of a RO filter made of multiple layers of porous and semipermeable membranes. Filtered water emerges from the center and the salty concentrate moves on through the outer layers. (Source: US Bureau of Reclamation Water Quality Improvement Center.)

Figure 9-9. An industrial scale desalination plant (Photo Courtesy of Siemens).

4. Deep Aquifers and Fossil Water

Aquifers at depths of tens to hundreds of meters underground have been highly utilized sources of freshwater for centuries. Household and farm water wells have tapped into these reserves. Recently, these shallow aquifers have been over-pumped and depleted in many parts of the world, marked by land subsidence, which is a noticeable sinking of large areas of ground, caused by the compaction of the depleted aquifer. It was not realized until very recently that huge aquifers exist at much greater depths. There are several recently discovered deep untapped aquifers around the world at depths of one to three kilometers underground, and there may be many more to be discovered. Some of the deep aquifers that are near the ocean may contain saltwater. Others may be fully sealed with no replenishing supply. The latter type could have been trapped under the Earth millions of years ago, and is thus often referred to as fossil water.

The Guarani Aquifer,[103] located beneath the surface of Argentina, Brazil, Paraguay, and Uruguay (Figure 9-10), is one of the largest deep aquifer systems in the world. It has an estimated thirty-seven thousand cubic kilometers of water deep underground at a maximum depth of 1,800 meters. It is not fossil water, as it gets naturally recharged at the rate of about 170 km³/year. By some estimates, it can supply the entire world's fresh drinking water for two hundred years if costs associated with water extraction and distribution networks were not of concern. The main known environmental side effect of extracting water from such aquifers is land subsidence, like what has been observed with the depletion of shallower aquifers. We can expect broader and more significant ground sinking with the depletion of such giant reservoirs.

Figure 9-10. The location of the giant Guarani aquifer (Source: Lily House-Peters).

---

[103] F. Sindicoa, R. Hiratab, A Manganelli, "The Guarani Aquifer System: from a beacon of hope to a question mark in the governance of transboundary aquifers," *Journal of Hydrology: Regional Studies*, volume 20, December 2018, pp. 49-59.

5. Large Scale Distribution Networks

Water distribution within a city or its vicinities has been practiced for millennia. For example, in the seventh century BC, Assyrians in the Middle East built an eighty-kilometer stone-lined canal that was twenty meters wide to bring fresh water to their capital of Nineveh. This included an impressive stone aqueduct that was over three hundred meters long.[104] Subterranean aqueducts or horizontal wells, often referred to as "qanat" or "kariz", were tens of kilometers long and brought free-flowing water from aquafers to cities or farms for consumption. They are marvels of ancient engineering (Figure 9-11), and many are still in use today.

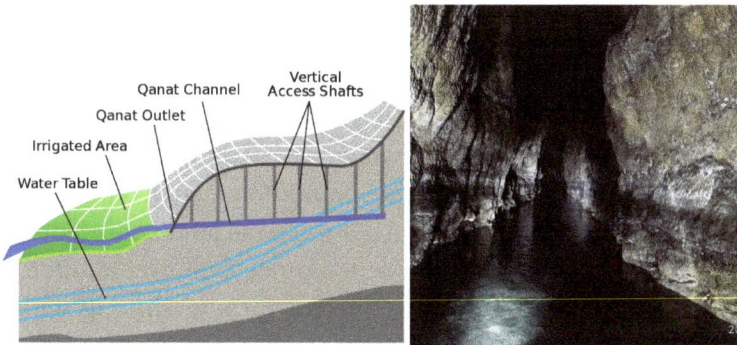

Figure 9-11. Kariz, of Gonabad subterranean water canal ("Qanats of Ghasabeh," Wikipedia) is over 25 centuries old, and it is still in use. The horizontal water canal is 45 km long and has 427 wells (access shafts). The deepest well, known as the mother well, is over 360 meters (about 1200 feet) deep.[105]

Most ancient aqueducts and canals, as well as most modern pipelines, were built to distribute water to districts and houses within or near a city. With today's growing urban centers overusing their local water supplies, it is becoming necessary to develop larger and longer distance networks to deliver water more equitably on a country-wide scale. These networks are essentially man-made rivers that run from water-rich areas to arid areas where fresh water is in short supply.

[104] Ernest Albert John Davies, "Canals and Inland Waterways," Encyclopedia Britannica.

[105] Babak Vaheddoost et al., "Estimating the Effect of Qanats and Underground Dam on Water Levels in Wells, Using Finite Difference Simulation.," ResearchGate, May 2014.

A recent example is the California Aqueduct Network, one branch of which carries water from Northern California to the Los Angeles area over a distance of approximately six hundred kilometers. Additionally, the massive South-North Water Transfer Project in China is a 2,400 kilometer network of canals and tunnels intended to transfer water from the Yangtze River in southern China to arid areas of northern China (Figure 9-12). Such projects are expected to become more massive over time, to the extent that local climates will be affected.

Figure 9-12. South–North Water Transfer Project Central route starting point Taocha in Xichuan County, Nanyang, Henan.[106]

6. Local Climate Control to Enhance Water Supply

Since the mid twentieth century, cloud seeding has been used with various degrees of success to stimulate clouds into dispensing their water at desired locations. Clouds consist of very small particles of water that are carried by air flow. Raindrops form when these water particles coalesce spontaneously or congregate around a nucleation site, such as a speck of dust. As they grow into larger droplets, they gain enough weight to fall to the ground. Cloud seeding adds more nucleation sites to the cloud in order to encourage this process and enhance precipitation. Cloud formation can also be aided by

---

[106] "Water Resources, China: Short-Term Fixes," Law-In-Action, October 8, 2014.

scattering chemicals in the air, such as calcium chloride, which is normally dispersed upwind from the target location for rain. These chemicals absorb invisible moisture from the air and form water particles that can start or enhance cloud formation. Once dense enough clouds are present, they may be sprayed with agents such as dry ice and silver iodide to initiate the condensation of water into larger droplets. Silver iodide is used because its molecular structure is very similar to ice. Ground-based generators as well as rockets, cannons, and airplanes have been used to deposit various types of agitation and nucleation particles to encourage both the formation and the growth of water droplets that eventually descend as rain (Figure 9-13).

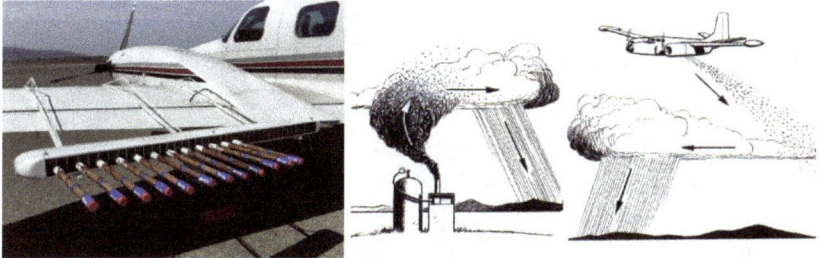

Figure 9-13. Right: Cloud-seeding schematics. Left: a photo of a plane carrying seeding flares. (San Luis Obispo County documents and "Creative Commons")

Cloud seeding is not capable of generating any significant amount of rain in areas where there would otherwise be no rainfall. Instead, it attempts to increase the amount of rainfall that a cloud would generate by itself. In other words, cloud seeding is only for rain enhancement, not desert rejuvenation. Since it is difficult to conduct controlled experiments on real clouds, there is no solid measured data to show the efficacy of cloud seeding. In other words, one cannot assert quantitatively how much rain a seeded cloud would have produced without seeding. However, the consensus based on simulations is that rain enhancement up to twenty percent can be achieved. If this assessment is correct, cloud seeding would be one of the least expensive methods of bringing fresh water to a region. The estimated cost of cloud seeding in California is twenty two dollars per acre-foot of water (1,233 cubic meters), compared with 1,890 dollars per acre-foot produced by sea water desalination.[107]

Cloud seeding is one of the emerging tools in the broader field of climate control. Since we are already affecting the climate with our industrial activities and our generation of atmospheric pollution, it is enticing to think that we may be able to steer climate change in our favor. It is known, for example, that a temporary cooling of global weather follows the spewing of a volcano's ash particles into the stratosphere. The reflection of sunlight back into space by volcanic particles is responsible for this. Perhaps similar cooling can be achieved by coating the clouds with dies to make them more reflective. Another known effect is that contrails generated by jet aircraft can grow into cirrus clouds. If enough of these clouds form above a region, they can cool the weather in the daytime and warm it in the nighttime. Engineers will eventually be able to simulate and use such effects to achieve climate control goals. Currently, one major hurdle, aside from technological capabilities and cost, is the "chaotic" nature of the climate.[108] Modifying the climate in one location can lead to large unintended changes in other locations. Such chaotic behavior cannot be modeled. This characteristic of the climate poses not only a big technical hurdle, but also a difficult geopolitical challenge.

Presently, climate control and water supply management are treated as disjointed issues, but they are inherently correlated, and solutions found for each need to consider the other. The management of water supply is also intertwined with food production and its distribution. Together, they will form the lifeblood of Simorgh that feeds its human nodes. A separate network of electrical power generation and transmission will feed the electronic and robotic nodes, as well as most tools and equipment that serve its multitude of functions. It can be seen that the support system for human nodes is significantly

---

[107] Brian Resnick, "Can California Make It Rain with Drones?" *The Atlantic*, October 7, 2014.

[108] A "chaotic" system is one that has high sensitivity to initial conditions, meaning that large effects can be caused by small perturbations.

more complex than the support system for electronic nodes. It is important for us humans to ensure the existence of a well established management network for favorable climate, water supply and food production by the early stages of Simorgh's development. This will mitigate any competition between humans and electronics for populating the nodes of Simorgh's body.

Summary

Despite the disturbing trends of global water shortage and climate change, and the availability of many techniques to fight them, relatively few substantial projects exist to address the issues.

Simorgh needs to be in control of its environment for its long-term survival. Human nodes are more vulnerable and need tighter environmental control than electronic nodes. Supply and distribution networks for water and food are of paramount importance for human nodes, while electronic nodes need electrical power distribution and less stringent temperature control. Therefore, it is important for us to build an effective infrastructure and establish good environmental practices early on so that supporting human nodes does not become a prohibitive liability for Simorgh.

Chapter 10: Past and Future of Jobs

For the majority of human history, most people worked seasonally or on an as-needed basis. In good economic times, shop owners had more regular job hours than others. Providers of daily goods and services, such as bakers and grocers, had a regular schedule year-round, while others like farmers and workers-for-hire experienced significant seasonal fluctuations.

Consistent work hours became more common with the industrial revolution. In fact, in the early years of mechanized factories, very long work hours were demanded by employers who wanted to maximize the outputs of their production lines. By the early twentieth century, labor unions, large companies, and legislation helped to standardize the forty-hour work week. Well-paid work in the manufacturing industry helped to establish a large middle class in many western countries. Employee loyalty to the employer was high, and large companies extended their umbrella of support not only to over the employees, but also over the communities in which they lived. Corporations supported schools, public works, and arts in the community. The thirty-year home mortgage structure that became common in the United States encouraged home ownership and created a path for employees to fully own their houses after thirty years of employment, ensuring a comfortable retirement. This employment model is rapidly changing, as short term, part time, and seasonal work are becoming more common, mainly due to automation and globalization. In a way we are reverting to the pre-industrialization work model.

Since the early days of industrialization through most of the twentieth century, machines and automation generally created more jobs than they replaced. This was due to the fact that automation was mostly used to augment manufacturing and production. More products could be fabricated per day and more crops could be produced per acre. Manual labor was replaced, but machines needed operators, and the increased productivity made more goods and merchandise available to the market. The increased volume of goods created a demand for additional services and distribution networks that employed a large number of relatively low-skilled workers. For example, the increased number of automobiles and trucks created a higher demand for roads, drivers, and service stations where low-skilled workers were in high demand. Another example is the growth of the airline industry which was made possible by faster aircraft production. It created a vast network of services such as airport operations, travel planning, ticketing, luggage handling, and so on, all of which increased low-skill employment.

> Automation of manufacturing created more jobs than it replaced. Automation of services does not.

Automation of manufacturing replaced some workers, but created more jobs in related services. Recent trends in automation now target the service industry itself. This type of automation generally does not result in a net gain in employment. Automation of services does increase job demand in other sectors such as information technology, but this is not an industry that employs a large number of low skilled workers. The resulting net change in the number of jobs is negative.

In the United States, approximately 12 million people[109] are employed

---

[109] United States Department of Transportation, Bureau of Transportation Statistics, "TET 2018 - Chapter 4 - Transportation Employment".

in the ground transportation industry as drivers of cars or trucks. Once these vehicles incorporate autonomous driving controlled by a central processor, there is no low-skilled substitution for human jobs lost. Similar trends are being seen in the airline industry, the food industry, and even in corporate services, such as accounting and office assistance. Below are some specifics on how automation affects some of the country's major job markets:

Ground and Air Transportation

The planning for personal driverless ground vehicles may have started with the concept of "smart highways" in the 1990s. It was thought at the time that electronic markers on the highways were needed to guide any autonomous vehicle. The smart highway concept was not limited to vehicle guidance and is still under development with a long list of functionalities including road hazard monitoring, photovoltaic power generation, inductive electric charging of moving vehicles, etc., but they are not essential for autonomous vehicle operation (Figure 10-1).

Currently, driverless automobiles guide themselves by cameras and sensors placed on the car itself and navigate with GPS. In the future, such vehicles will be able to communicate with each other on the road and with any remaining traffic control signals. This will make automated cars much safer than human-driven cars. In fact, it is likely that human-controlled vehicles may be outlawed in certain municipalities.

Driverless trucks won't need to stop for rest, and being able to control and manage a fleet of trucks on a remote screen is very efficient for trucking companies. Driverless trucks can be used at locations such as mining sites where conditions may not be healthy for human drivers. Semi-autonomous trucks are currently on the road. The number of fully autonomous trucks is expected to become significant by 2024 and grow at an annual rate of over 25% over the next few years.[110]

---

[110] "Semi-Autonomous & Autonomous Truck Market" 2020 report by Markets and Markets.

Figure 10-1. Smart roads and highways may have smart pavements (left) with inductive charging lanes for electric cars, solar electricity generation, and heaters for melting snow and ice. Other features include cameras and displays for traffic management, and road hazard warning systems.

Driverless taxis and shared cars that can be rented on demand are on the rise. This will further reduce the market size for professional drivers. There are approximately twelve million people in the U.S. who earn their living by driving vehicles. It is clear that most of these jobs will dwindle without a replacement. The more enduring jobs in this sector will likely to be in corporate management, marketing, resource allocation, as well as in developing and maintaining control algorithms.

Jobs in air transportation include the flying crew and the ground support such as air traffic control and airport services. Routine air traffic control and air space management are currently performed by a combination of software algorithms based on flight plans, radar and GPS information, and human oversight (Figure 14-2). There are approximately fourteen thousand air traffic controllers employed in the U.S., and the number is expected to remain relatively constant within the next decade.[111] Traditionally any instructions given by the airport tower to the aircraft was conveyed by a dialogue between the pilot and the air traffic controller. Increased traffic and the addition of other flying objects such as drones and air taxis are forcing this

---

[111] U.S. Department of Transportation and Federal Aviation Administration, A Plan for the Future: 2006-2015: The Federal Aviation ADMINISTRATION'S 10 Year Strategy for the Air Traffic CONTROL Workforce (Washington, D.C., VA: Federal Aviation Administration, 2016).

communication to become digital. Air traffic controllers are not necessarily as focused on personally guiding every single flight from route and approach through landing as they were in the past. Instead, their main role is to manage traffic by intervening and overwriting the automation when something unexpected happens. Currently, automation of air traffic control serves to assist human controllers and to allow more flights to be managed by each controller. Even though the eventual full automation of air traffic control will only enhance the efficiency and safety of the system, this profession's replacement with automation will be relatively slow, partly due to liability concerns.

As far as the flying crew is concerned, airplanes operate on autopilot for most of the duration of the flight. Takeoff and landing are still performed manually in many cases, except in times of poor visibility. In a fly-by-wire airplane, all functions are performed by remote control through a graphical user interface. The commands may be given one at a time by the pilot, or they may be delivered sequentially by an automation algorithm. Instead of setting individual control levers of the plane, the pilot issues higher level commands such as "go to a different altitude" or "descend at a given rate," This makes flying a lot easier for the pilot. The cockpits of passenger airplanes used to seat three operators: the pilot, the copilot, and the flight engineer.[112] Automation made the post of the flight engineer unnecessary. Currently, manufacturers are developing one-person cockpits for passenger airplanes, eliminating the role of the copilot. It is only a matter of time until autopilot will handle the entire flight in commercial air travel. Autonomous air taxis that are already operational in select locations around the world will slowly increase people's confidence in pilotless planes. But for reasons of liability, the phase-out of commercial airline pilot jobs will be a slow process that will take several decades. Food service in the cabin is also benefiting from automation. Both the preparation of food trays and their distribution within the cabin will become more automated with time. Only a skeleton crew will be needed in pilotless planes for

---

[112] Earlier cockpits had even more specialists such as a navigator and a radio operator.

minor routine tasks and emergency response.

Many airport services, such as passenger check-in and baggage handling, have become automated. The lingering role of humans in the air transportation industry is likely to be in corporate management of the airline companies. Business decisions such as resource allocation and market positioning can and perhaps should be reserved for Simorgh's human nodes.

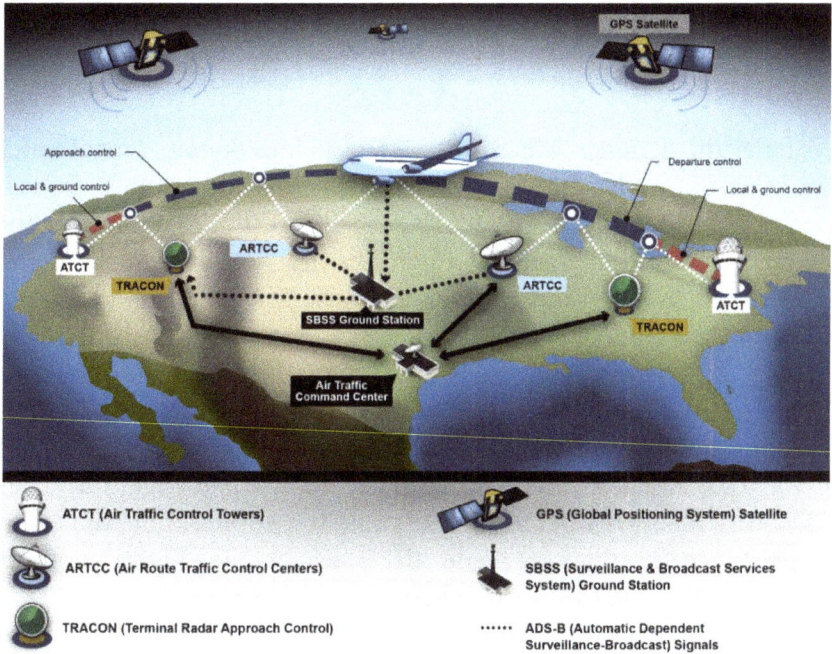

Figure 10-2. Information resources used by air traffic controllers.[113]

Manufacturing

Manufacturing normally refers to the fabrication of identical products in large quantities. The process involves repetitive tasks that often require some precision. A manufactured item is identified by a model-number that is expected to represent the full set of

---

[113] David Kravets, "US air traffic control computer system vulnerable to terrorist hackers", arstechnica.com, March 2015.

specifications for the unit. Different items with the same part number need to have the same "form, fit, and function." Therefore, on the manufacturing floor, all processes and materials should remain the same, day in and day out. Humans performing fabrication tasks at many such factories tend to suffer from boredom and fatigue. A programmed robot on the other hand is much more suited for the fabrication of identical products. A robot can work around the clock without a break, and the quality of its work will not suffer from exhaustion. Let's examine more closely two areas of manufacturing that have high visibility to the general public: electronic device fabrication and automobile assembly.

At the core of many electronic devices are integrated circuits (ICs) that are typically fabricated on a semiconductor substrate called a wafer.[114] Integrated circuits perform many functions such as wireless reception, data storage, and programmed calculations. IC manufacturing is done in semiconductor fabrication facilities (fabs) that transfer multilayer intricate patterns from "photomasks" to the semiconductor material. The patterns have very small dimensions and could be blocked by any speck of dust or a small particle that may land on the wafer prior to or during the process. In order to avoid such problems, the processing is done in controlled environments known as "clean rooms" that use filtering to limit the number of particles per unit volume of air in the room. Operators who enter the clean room need to wear special overgarments that prevent particles from being released into the air from the operator's hair, breath, or clothing. People become accustomed to wearing such garments, but they are not a comfortable or natural outfit. Modern fabs operate with very little human labor (Figure 10-3). Instead, wafers are transported robotically from one process chamber to another, and completed wafers go through automated testing and dicing before final delivery. This way the fab can operate like a single automated machine. Human involvement is expected to continue in the design

---

[114] Semiconductors are a class of materials known for their control of electric current flow. The most commonly used semiconductor material is silicon on which many computer chips are fabricated. There are also compound semiconductors such as gallium nitride that are good emitters of light and are used in the fabrication of light emitting diodes (LEDs).

of both the integrated circuits and the fabrication machines that are used in the fab.

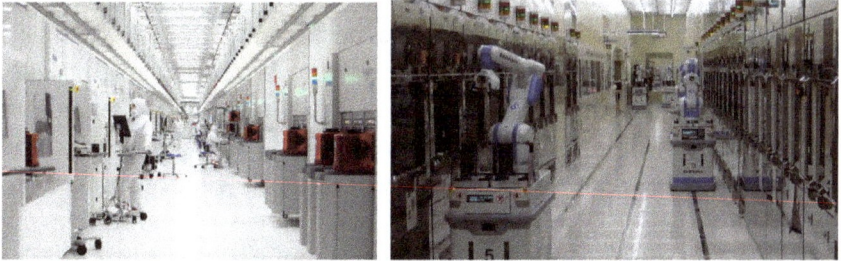

Figure 10-3. People who work in a semiconductor fabrication facility need to wear special clothing to avoid spewing particles in the air (Left Image, Intel-Micron fab). Robots are cleaner than humans in gowns (Right Image, Samsung fab).

Automobile manufacturing initially started in plants or shops, where everyone worked on completing the assembly of one automobile before moving to the next. Assembly lines pioneered by the Ford Motor Company changed this process. People stayed at one location while automobiles moved from one assembly area to another. Every worker repeated the same task on successive automobiles. As a result, manufacturing efficiency increased significantly, and training for each position in the assembly line became much simpler because of the limited responsibilities of each employee. Specialized tools for lifting, cutting, and welding were developed to make the job of each worker much easier. When automation was introduced into these specialized tools, the operators' job was reduced to positioning and pressing the buttons on the machine. Finally, the tools became robots that are now controlled by software algorithms instead of humans (Figure 10-4). Modern automobile manufacturing employs far fewer workers per delivered vehicle than the original assembly lines.

The automation of manufacturing leads to higher productivity that in turn promotes additional employment in its support industries. Therefore, this type of automation does not lead to a net job loss in the overall economy. In the manufacturing industry itself, the lingering jobs for humans include: robotics and machine learning specialties, as well as marketing, and sales personnel.

Figure 10-4. Automobile assembly line. The whole factory may be considered one automobile making robot that employs a few human monitors. Source: BMW/Fred Rollison.

Maintenance and Repair Service

Repairing of many items such as household appliances and electronic gadgets is no longer common practice. The full replacement of the faulty unit is the more attractive option. Replacement has become the preferred method due to the relatively high cost of repair compared to the relatively low manufacturing cost of a new item. Additionally, the quick obsolescence of old technology favors purchasing a more up-to-date replacement than opting for repair. Consequently, most of what were once private electronic repair shops are no longer in business. Routine maintenance and repair are only done on high-ticket price items such as automobiles, aircraft, and heavy machinery used in factories, and in large projects.

The trend in the design of highly priced equipment is to incorporate internal fault detection and automatic diagnostics in order to communicate any problem with the user. In automobiles, this is called OBD, or "onboard diagnostics." It was originally introduced as an aid to repair personnel, but it has now become accessible directly to the owners through smart phone applications. It is envisioned that most future equipment will be modularized and the OBD will determine which module needs to be replaced. This will significantly reduce the amount of human labor required to complete the repair. Eventually, what is left of the diagnosis and repair process will become automated, and parts replacements will be performed by specialized robots.

Lingering human jobs in maintenance and repair industry are specialties in software, sensors, and robotics.

Construction

The trend in large construction projects has been the incorporation of prefabricated sections or modules that are brought to the construction site for assembly into a building. This is also known as remote fabrication. The prefabricated sections may be entire walls, including windows, doors, and lighting fixtures, or entire bathrooms or even complete small apartments. Large industrial buildings can be fully designed by 3D modeling using prefabricate modules. 3D modeling allows the inclusion of all wiring, plumbing, and HVAC ducts into the design. Building sections that are replicated many times are ideal candidates for prefabrication. (Fig 10-5a). Building designers benefit from using standard modules because such modules become optimized and streamlined over time as a result of repeated use. Remote fabrication sidesteps the need to adapt to climate and weather variabilities on the construction site, which could disrupt work schedules. Manufacturing equipment is kept at a centralized location in a controlled environment where it can operate day and night.

There is another technology on the horizon for buildings that don't lend themselves to modular construction. The concept of 3D printing that started with the fabrication of solid models for small artifacts is rapidly advancing, and some companies are testing the idea of 3D printing entire buildings, with wiring and plumbing automatically inserted during construction (Figure 10-5b). This is a relatively slow process, but it will allow for shapes and features that would not be possible with traditional construction technologies.

(a)   (b)

Figure 10-5. Left: Modular construction is done by assembling prefabricated units into a complete building (MMC Quantity Surveyors). Right: A construction 3D printer prototype (https://www.designingbuildings.co.uk/wiki/3D_printing_in_construction).

According to the Bureau of Labor Statistics, there are over seven million construction-related jobs in the U.S., growing at approximately one percent per year. Table 10-1 shows a subset of this number as occupations related to building construction. The total number is expected to grow by twelve percent over a ten-year period. Automation will transform this field from a tradesmen's market into a manufacturing industry. Remote fabrication will become prevalent, and Lego style modular building assembly will dominate apartment complex and office building projects. Onsite manual labor will be augmented by exoskeletons and robots. Inspection and quality assurance will be conducted by remote monitoring enhanced by drones and avatars.

Occupation	2016	2026	Future Advancements
Carpenters	329,400	357,800	Prefab
Construction laborers	227,300	259,600	Exoskeletons, Robots
Construction managers	99,200	113,100	Remote monitoring, Drones
Cost estimators	39,500	45,100	Standardization
Cement masons and concrete finishers	23,900	27,200	Prefab
Painters, and maintenance workers	22,000	25,200	Robots, Sensors
Civil engineers	19,900	22,600	Virtual Reality
Helpers	18,200	20,800	Exoskeletons, Robots
Structural iron and steel workers	15,800	17,900	Prefab, Robotic Arms
Drywall and ceiling tile installers	10,500	12,000	Prefab, Exoskeletons, Robots
Total	805,700	901,300	Decline in on-site jobs

Table 10-1. U.S. occupations in building construction in 2016 (actual) and 2026 (projected). The right column lists the advancement trends in each occupation. Source: U.S. Bureau of Labor Statistics, Office of Occupational Statistics and Employment Projections.

Agriculture

The mechanization of agriculture or power farming began with the advent of the tractor. Steam powered engines that kickstarted the Industrial Revolution began to be used in some farms in the late eighteenth century to help with threshing wheat. Steam engines were large and difficult to maneuver, with the potential hazard to set the grain on fire. The first gasoline powered tractor that was safer and easier to use was introduced in late-nineteenth century by John Froelich (Figure 10-6).

Figure 10-6. An early Froelich tractor. This brand was the first gasoline-powered tractor.

Source: farmcollector.com

Mechanization, combined with better fertilization, naturally increased crop yield per farm and per farm worker. Gradually, fewer and fewer people were needed to work on the farms. Figure (10-7) shows that while traditionally sixty to seventy percent of a country's workforce was once employed in agriculture, this ratio has dropped to less than five percent in many countries today, even though the agricultural output is significantly higher now.

Recently, the development emphasis in farming has shifted from mechanization to farm automation such as the introduction of farm robots (Figure 14-8), and genetic refinement of seeds for higher productivity. This combination will further reduce the number of low-skill jobs in agriculture as production levels continue to rise.

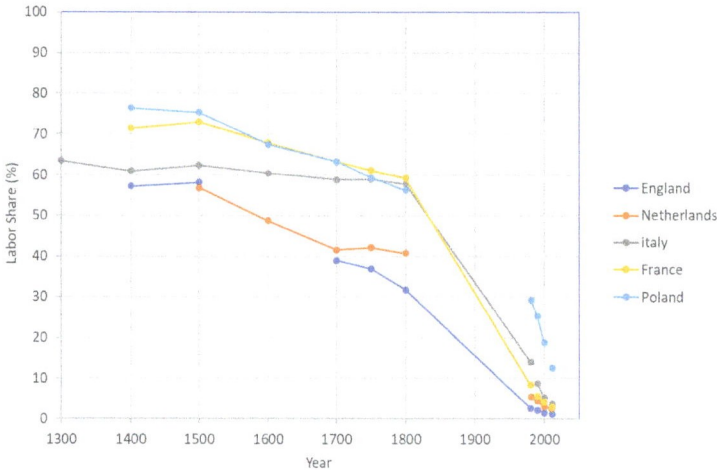

Figure 10-7. Share of labor force working in agriculture in some European Countries, since 1300 – by Max Roser.[115]

It is noteworthy that this is another production industry whose increased efficiency due to automation does not result in a net job loss in the economy. Other new occupations are created in food preparation and distribution due to the abundance of farm products. A net job loss ensues only from the automation of service-related tasks that do not directly add products to the market.

---

[115] Max Roser, "Employment in Agriculture," Our World in Data, April 26, 2013.

Figure 10-8. An autonomous tractor (CNH Industrial) and a farming robot that performs weeding and crop maintenance (Deepfield Robotics).

Another recent movement in agriculture is the development of synthetic food and cultured meat that is likely to reduce people's reliance on livestock farming. The satirical concept of a meal in a pill,[116] as depicted in the old cartoon of Figure 10-9 is of course not a viable alternative to today's food consumption. Eating is more than just providing sustenance for the body; it is also an activity to be thoroughly enjoyed. Any advancement in food production technology has to take this fact into account. It is worth noting however that people's tastes in food change over time, and dishes that may have been highly sought after a century ago may not be very appetizing today. For example, the broad desire for low-fat foods was sparked by health concerns at first, but as a result of its widespread adoption, many people's palates have become intolerant of fatty foods over time. Additionally, "new world" food items that did not exist in Europe and Asia before the discovery of the Americas are now an indispensable part of most people's diets around the world.[117]

---

[116] The concept of a meal in a pill is an exaggeration that became a topic of discussion in the mid-twentieth century with the inception of space flight and the development of compact and portable "astronaut food." The appeal of a meal in a small package persists only in limited applications, such as when easy distribution is needed for the reduction of malnutrition in certain areas of the world.

[117] Popular examples of "new world" food include corn, potato, tomato, vanilla, and chocolate.

Figure 10-9. Thanksgiving meal in a pill (September 19, 1926, Ogden Standard-Examiner).

Livestock farms are, objectively speaking, protein factories for humans, albeit inefficient ones, as they consume a lot of energy and produce much waste. Traditional livestock farming is also very wasteful of land and water per calorie of food produced. In addition, animal rights activists believe that livestock farming practices are cruel to animals.

Farm animals have been bred for human consumption and farm profit over the centuries. Special breeds of cattle can produce up to twice as much milk or have lean muscle with low fat content driven by market demand. Recently, human genes have been successfully introduced into dairy cows by genetic engineering, allowing them to produce milk that is more similar to human milk. Other developments strive to produce meat with certain medical benefits.[118] Despite these advances, the low efficiency of livestock farming in protein production remains a concern.

The term "protein efficiency" refers to a farm animal's ability to convert its food into protein in the resulting farm product (Figure 10-10). Protein production is most efficient with eggs at twenty-five percent, followed by milk production, while beef production is quite inefficient. Additionally, livestock farming produces significant air pollution, and most of the products that are meant for human

---

[118] Natalie Wolchover, "Cows Make Humanized Milk. But Is It Safe?", livescience.com, June 10, 2011.

consumption are not inherently sterile.

This provides impetus for the development of alternative protein production methods. It is true that plants can supply various sources of protein, and people may choose a vegetarian lifestyle. But most people still prefer not to eliminate all meat and dairy products from their diet. To this end, there are meat substitutes made from vegetables, and there is the emerging technology of cultured meat.

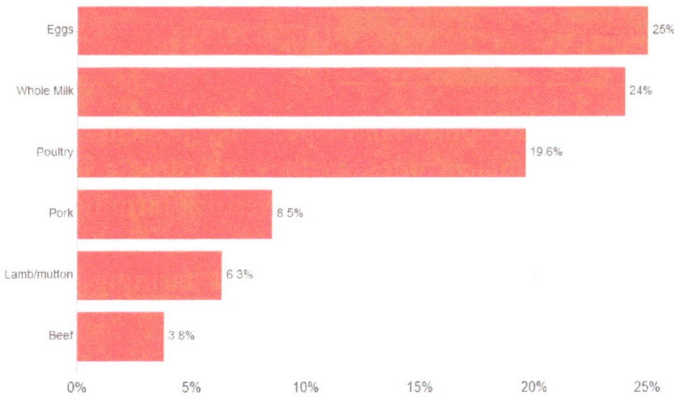

Figure 10-10. Protein efficiency of meat and dairy products. Protein efficiency is the percentage of protein in the animal feed that gets converted to protein in the farm product.[119]

Cultured meat, or "clean meat," is made by growing muscle cells in a nutrient environment and forming them into muscle-like fibers (Figure 10-11). This field of research is sometimes referred to as "cellular agriculture." Cultured meat is potentially healthier for human consumption than regular meat since it may be engineered to include healthier fats and no pathogens. It can also be grown with a larger variety of tastes that may or may not exist in nature. For example, the cultured meat may be made to taste similar to the meat of certain rare animals that are considered delicacies.

> Food distribution will become a major body function of Simorgh.

---

[119] Hannah Ritchie and Max Roser, "Meat and Dairy Production," Our World in Data, August 25, 2017.

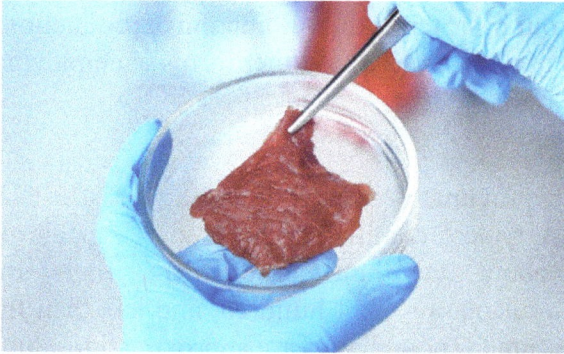

Figure 10-11. Proof of concept of cultured meat. It is still a curiosity item and not close to production. Source: New Food Magazine, 12 March 2018.

The concept of having credible alternatives to animal meat has far-reaching potential. The growth of human population and rising incomes has amplified the demand for meat, whose production competes with crops for land space. Cultured meat and dairy product alternatives can reduce the demand for wasteful livestock farming to free up more agricultural land while simultaneously supplying consumers with more diverse, healthy and sterile food. Currently, the technology of cultured meat is in its infancy, and the price of the resulting products is extremely high. Once the cost of cultured meat becomes comparable with organic meat, its adoption will undoubtably accelerate.

Cultured meat may have a role as a gap filler during the time it takes for people's taste in food to turn away from animal product consumption. As new types of proteins are synthesized and tasty recipes evolve around them, people may lose interest in consuming products made by living cells.

It is evident that livestock farming is poised for size reduction, causing the associated low-skill jobs to vanish. However, food production and distribution as a whole will continue to employ a large number of people. In fact, food distribution will become a major body function of Simorgh and its lifeblood for human cells.

To reiterate our basic tenet, the automation of production, unlike the automation of services, leads to other forms of employment.

Healthcare

There are approximately thirty thousand diseases identified for humans[120]. A typical community hospital is capable of routinely diagnosing about one thousand of them. There are standard treatments available for eight hundred, and less than five hundred diseases are cured. These figures are approximations, but they allow for some broad conclusions to be drawn. Senescence diseases, as well as "minor" and "uncommon" diseases, are not likely to receive effective treatment in a community hospital, but the good news is that the treated and cured diseases include some of deadliest and most encountered. Nevertheless, if you happen to have a rare disease, the chances of an average physician making an accurate and timely diagnosis are slim.

Traditionally, a physician's ability to make the correct diagnosis was based on reference books in his possession and prior experience with patients, neither of which provided immediate help for uncommon diseases. But consider an intelligent computer database that can search the entire collection of human diseases based on the patient's symptoms and quickly provide a list of candidate diseases with their associated probabilities that the patient may be suffering from. Such systems, called Computer Aided Diagnosis (or CADx) already exist at various degrees of sophistication. CADx makes it much easier for your physician to arrive at the correct diagnosis with the most up-to-date treatment.

The development of medical diagnostic algorithms employing expert systems and artificial intelligence techniques started in the early 1970s, with one of the first being Stanford University's Mycin system for the identification of bacteria. These systems have been mostly limited

---

[120] Numbers from 10,000 to 70,000 have been quoted. International Statistical Classification of Diseases and Related Health Problems (ICD-10) has 70,000 codes for human diseases. "Faster Cures" affiliated with the Milken Institute (http://www.fastercures.org) mentions "10,000" diseases. NIH and some European sources set the figure at 30,000.

to particular areas of diagnosis. For example, there are algorithms actively in use for reading mammograms or electrocardiograms. More recently, there have been efforts to develop widely applicable CADx systems that would consolidate a large amount of data and generate a highly personalized diagnosis for each patient. Examples include Ginko CADx and IBM's introduction of its machine learning platform Watson to healthcare diagnosis (Figure 10-12).

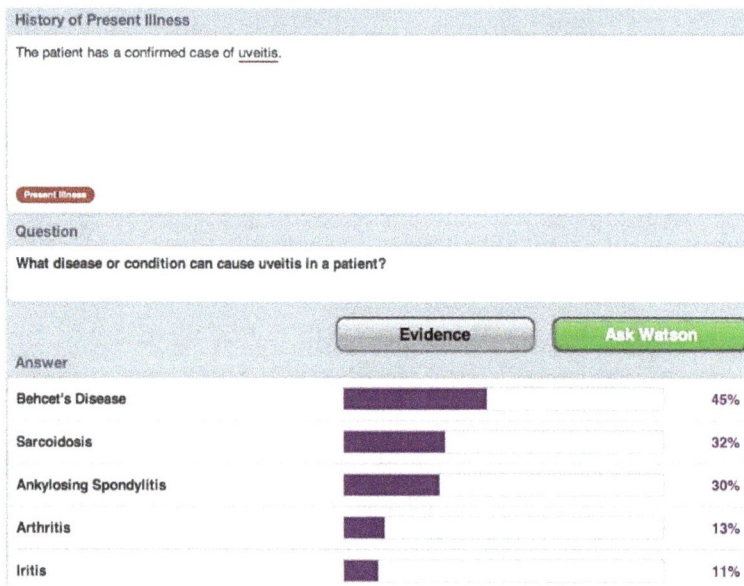

History of Present Illness

The patient has a confirmed case of uveitis.

Present Illness

Question

What disease or condition can cause uveitis in a patient?

Evidence | Ask Watson

Answer

Behcet's Disease	45%
Sarcoidosis	32%
Ankylosing Spondylitis	30%
Arthritis	13%
Iritis	11%

Figure 10-12. An example of how a CADx system that could answer a particular medical question based on the patient's background information and the available medical data (Watson Health).

In addition to diagnosis, automation is helping other areas of healthcare, including pharmacy and surgery. Surgical robots are essential in most "minimally invasive" procedures, and they are taking strides toward the promising field of remote surgery.

Robots routinely carry out the crucial steps in some procedures, including orthopedic knee replacements, vision correction eye surgery, and hair transplants. What these types of surgery have in common is the rigid nature of their targets. Leg bones, eyes, and heads can be held firmly in place during the procedure. Performing surgery on soft tissue is more challenging to automate, because soft tissue cannot be immobilized, and its movement is hard to track. The da Vinci system by Intuitive Surgical is being used for soft tissue surgery, but it currently has no automation. Its arms are equipped with precision surgical tools and imaging cameras that can be moved with high precision. The system is teleoperated by a surgeon enabled with live video and tactile feedback.[121] As data networks become faster and more reliable, such surgical procedures may be performed remotely.

An autonomous surgical system by the name of STAR[122] is capable of performing interstitial suturing of soft tissue automatically with higher precision and repeatability than human hands. It tackles the soft tissue challenge by combining multiple fluorescent and 3D imaging. It automatically prepares a suturing plan and revises it during the procedure based on the observed soft tissue movement.[123]

Automation is revolutionizing healthcare. It will eventually be unnecessary for most patients to visit a physician's office or a hospital. Portable sensors and interactive diagnostic algorithms will perform both preventive care and identify the cause of any ailment. Treatment can be administered with minimal human involvement. Most healthcare service jobs will be replaced with automation. Specialized jobs in the areas of psychology and mental health will persist longer in their current form, though advances in automated counseling and behavioral neuroscience are likely to change the nature of this discipline as well.

Retail

Two major approaches to shopping are emerging: You can go to a store, pick up the items you want, and walk out. You will be automatically charged and the funds will be taken out of your account. The other way is to order online and have the items delivered to you the same

---

[121] Tactile feedback allows the surgeon to feel the softness or hardness of the tissue during remote surgery.

[122] Developed by the Children's National Health System in Washington, D.C.

[123] IEEE Spectrum, May 2016.

day or within hours. Online shopping will be made more appealing by small brick and mortar show rooms where you can examine merchandise up close to help you with your shopping choice before you order online. Both methods may eventually converge with the use of virtual reality. You will be able to closely examine the item remotely in a virtual reality setting. For example, you can experience walking into a clothing store, selecting a garment, seeing yourself wearing it, and finally choosing to own it, all without physically leaving your workstation.

With retail moving in this direction, one can see that cashiers' positions in stores are already being eliminated in favor of self-checkouts that will also become obsolete in the near future, as invisible machines will identify the items that you leave the store with. Examples can be seen in "Amazon Go" stores (Figure 10-13). As shelves are automatically replenished, and items are automatically ordered to resupply the inventory, human attendees will only have supervisory roles to ensure that everything operates properly. For instance, a ten thousand-square-foot grocery store ($1,000$ m^2) may be run by as few as three employees per shift. More human involvement will be needed for non-repetitive tasks such as designing store layouts and opening new stores or dismantling old ones.

Figure 10-13. The "Just Walk Out" shopping experience at "Amazon Go."[124]

---

[124] Pheobe Tran, "Amazon Go to Open Six More Locations, ALBERTSONS Buys Rite Aid as Amazon Threat LOOMS + More," Food+Tech Connect, April 10, 2018.

In the year 2016, a total of 4.6 million people in the United States were employed as cashiers[125]. This has been one of the most popular "low skill" jobs in our economy. Cashier employment in the United States is expected to fall by ten percent between 2020 and 2030. Inventory and warehouse management are also becoming fully automated, and humans who work in those positions are beginning to rely heavily on robotic help for moving and repositioning warehoused items. Items that have predetermined shapes and weights such as boxes and identical items are very easily handled by machines. Grabbing and placing random objects such as a basketball followed by a pair of socks can still be challenging to robots, but their skills are becoming more refined with the help of new AI techniques.

Human work will still be needed in management, planning, and technology supervision in the retail industry.

Personal Banking and Financial Services

Traditionally, a bank was a secure place for people to keep their money and secure loans when they needed them. It was a physical location in the neighborhood, and the bankers knew the majority of their regular customers. Over the years, banks grew in size, and the way they interacted with their customers changed. The customer was identified through a bank account number, and their identity was verified by pictures, handwritten signatures, and personal data such as birth date, family history, and so on. Credit cards were issued that were physically difficult to duplicate, and the owner could use them for credit on demand (Figure 10-14). This state of banking affairs survived for a few decades before starting to change rapidly. Now, customers hardly ever go to the bank. More commonly, people know their bank merely through an application on their mobile phones. From the perspective of such customers, the bank is primarily a ledger that keeps track of deposits and credits versus expenses. Credit worthiness is an index calculated automatically based on one's

---

[125] United Stated Department of Labor, Bureau of Labor Statistics: www.bls.gov/ooh/sales/cashiers.htm

transaction history, and credit cards are no longer necessary. Online payment apps can manage a person's credit without the assistance of a bank employee.

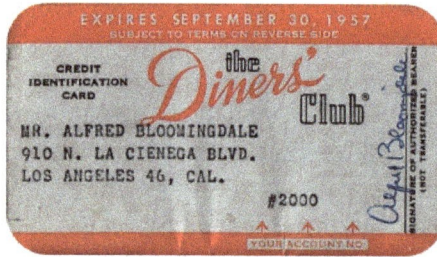

Figure 10-14. An early Diners' Club credit card. The first independent credit card company, Diners' Club, was formed in 1950. Source: Smithsonian National Museum of American History.

Another service that banks and brokerage houses provided for their customers was to offer strategies for savings and investment. This was done by financial consultants at the bank. The customer would explain his financial outlook and future needs, such as college funds and retirement plans, and the consultant would recommend monthly savings and investment plans accordingly. All of these financial services are likely to be provided in the near future by a "smart bank account," which is a bank account that recognizes you based on biometric information. It understands your resources and goals, and it plans your investments for you. It can extend credit to you when you need it, and it can advise you whether or not to proceed with a particular discretionary spending, as well as warn you if the discretionary spending is beyond your means. It will start by acting as your personal advisor, but as technology advances, it will be able take care of most routine purchases and spending for you. It will manage your income, your expenses, your savings, and your credit according to your stated preferences. You will only interact with your smart bank account, not with your bank. The bank will be an algorithm that runs in the background.

Money is also going through a transformation. Since the dawn of humanity, it has been necessary for people to exchange excess goods and crafts for food and materials they need. This exchange was done directly by bartering without any independent measure of value for the exchanged commodities. When needed, rudimentary units of common items such as cattle or grain were used as measures of value. One bundle of goods would be priced at half a sheep for instance and another bundle for two sheep. Shell money appeared in the second millennium BC. Shells or pieces of shells that were initially worn as body ornaments became convenient intermediaries and value assignment tools for buying and selling goods.

The first known precious-metal minted coins were made of silver and gold in the nation of Lydia in Asia Minor (present day Turkey) around 600 BC. These early minted coins had the stamped image of a lion (Figure 10-15). Other nations followed suit, and silver and gold minted coins remained the most commonly used forms of money for centuries. Bank notes and government issued paper money didn't come into full circulation until the seventeenth century AD, even though promissory notes, on paper and leather, did have limited circulation much earlier.

Figure 10-15. Lydian coin made of an alloy of gold and silver. Source: British Museum.

Recently, direct use of money has been mostly replaced with electronic transactions, and the name of the currency, such as the U.S. dollar, is only used as a metric for value. Financial policies adopted by each government affect the valuation of their currency against others. Deficit spending for example, may cause inflation and a gradual currency devaluation. This works to the disadvantage of holders of the currency. It can be envisioned that in the future, the use of global cryptocurrencies such as bitcoin will become more attractive to the user, despite initial opposition from governments and central banks. The use of a universal currency has many advantages such as bypassing currency conversion uncertainties in global trade and serving as a unified global metric for the valuation of goods and services. Some countries will object to this concept because it restricts their ability to use independent monetary policy, such as deficit spending and controlling the money supply, to stimulate or moderate the economy.

> The best form factor for any robot to fulfill a particular job is hardly ever a human shape, unless it is intended to mimic a human, use human tools, or the person who interacts with it prefers that shape.

Major banking services that people have relied on are: (1) The storing of value for the owner; (2) The moving of value from one owner to another; (3) Making credit available; (4) Providing financial planning and advice. It is clear that none of these services require the existence of a physical bank. We can expect that banking services will be provided by Fintech (Financial Technology) algorithms and virtual companies. Human employment in this sector will be limited to personnel who decide and implement structural changes and updates to the service menu.

Home Services

We refer to home services as work done in private residences for regular upkeep and maintenance. These services include house cleaning, decorating, and gardening, as well as repair and renovation services such as plumbing, HVAC, electrical work, painting, roofing, and flooring. Traditionally, such services were performed by permanent staff in upper class homes and mansions, by the occupants themselves in lower class dwellings, and by local shops and craftsmen for the rest of the population. Currently, there are about nine million people employed in this sector of the economy in the United States.[126]

There have been some technological advances that have simplified cleaning tasks. Other tasks – particularly those that include repair work – have not gone through any significant change yet, even though there are tools that simplify the job of the repairman. As the reliability of home construction improves and more types of appliances and fixtures are modularized, it is envisioned that repair services will become less and less needed and defective parts are replaced instead of repaired. Rapid obsolescence is another factor that favors replacement over repair.

One temptation that must be resisted is to think that humanoid robots will be performing many of the listed home services in the future. This is legacy thinking that is often upstaged by a paradigm shift. When automation uses a "basic principles"[127] method of design, the outcome may not have any similarity to the existing practice. With basic principles design, the appearance of the machine will be dictated by the job to be performed. Figure 10-16 shows how automatic vacuum cleaners were initially imagined and how they actually evolved.

Figure 10-16. Cleaning robots as imagined in the past (Javier Pierini /Taxi/ Getty Images), and as they actually appeared on the market (iRobot, and Dyson, existing cleaners).

---

[126] "Other Services (except Public Administration) in the US," IBISWorld, accessed August 19, 2021.

[127] Basic principles method is in contrast with evolutionary method. Basic principles ignore current practice, and designs based on the needed functionality.

Humanoid robots will be primarily used as companions or assistants to humans because we can identify with them or even bond with them; but Simorgh doesn't need humanoid robots. Simorgh initially forms with an excess of under-skilled humans that it will prefer to do without in favor of robots. The best form factor for any robot to fulfill a particular job is hardly ever a human shape, unless it is intended to mimic a human, use human tools, or the person who interacts with it prefers that shape.

Low-skill employment in home services is on a downward trend. Jobs with more longevity include home decoration and landscaping, which entail an element of art even though they also take advantage of computer assistance for details and dimensions.

Entertainment (Motion Picture)

CGI stands for Computer Generated Imagery, which is the latest successor to animated cartoons. In the entertainment industry, cartoons and animations initially started as 2D hand drawings that were brought to life by sequential imaging. They looked much different from standard movies, and naturally created their own separate category in film making. Hand-drawn figures were later replaced by computer-generated graphics and animation that led to much faster production with less manual labor. Recently, computer generated imagery has started to create more realistic animated characters with detailed skin features and natural looking movements that resemble live artists. Current CGI-generated actors are not easily differentiated from live performers. In fact, some have been given personalities that people identify with and may be used in place of actors (Figure 10-17).

Additionally, many stunts in movies that were once performed by highly talented personnel are now being done by CGI. Computer-generated actors may be completely imaginary, or copies of actual people. It is envisioned that if an actor is filmed for a short period of time, enough information will be collected to allow an entire movie to be completed by CGI without the actor's participation. This will reduce the demand for actors and all the logistical support needed to record live acting scenes.

Set construction delivers a film's scenery and provides the location details for movie production. Set construction traditionally took a long time and required a large budget, depending on the subject matter of the movie. For example, the movie *Cleopatra* was initially planned to be filmed in England, but the original set that was built for it was destroyed in a winter storm. Subsequently, they moved the filming location to Rome, where a set was built at such a grand scale that reportedly, the builders faced a building material shortage (Figure 10-18). Motion picture technology is currently moving away from elaborate set construction and toward montage of acting over natural scenery or computer-generated structures.

Figure 10-17. Japanese computer-generated character "Saya" can play in movies.[128]

---

[128] Juan Buis, "This Girl Isn't Real, And It's Proof That CGI Isn't Creepy Anymore," TNW | Creativity, April 27, 2021.

Figure 10-18. The elaborate movie set constructed for the motion picture "Cleopatra" in 1963. Photo by Silver Screen Collection/Getty Images.

Low-skilled jobs such as movie production support staff and set construction workers are becoming obsolete. There will be continued employment in CGI. Acting in both movies and theatre will continue, although in competition with CGI.

Entertainment (Music Production)

Before modern civilization, music started with singing that could be as simple as a lullaby. For a larger audience, songs could be accompanied by drumming on various objects, clapping, or stomping. Later, musicians started to build improvised instruments for making pleasant sounds. These included early drums and wind instruments, such as whistles and flutes made of bones and bamboo. String instruments possibly started by making sound with one's hunting bow and arrow. Formal development and standardization of musical instruments are believed to have originated in Mesopotamia about twenty-five to thirty centuries ago.[129] The courts of the Sassanian kings of Persia were renowned for their lavish musical entertainment, and they developed a major business of exporting musical instruments

---

[129] A. Percival, *History of Music*, The English Universities, Ltd. 1961.

(Figure 10-19). Famous Greek mathematician Archimedes is credited for pioneering the designation of musical notes and the beginning of musical scales. He went as far as suggesting that the entire universe was being played as a musical instrument.[130]

For many centuries, people's exposure to music was limited to listening to the live performance of someone they knew, or to neighborhood musicians who would play at social or religious occasions. Higher quality music could be enjoyed in a concert hall, but not very often for the average person. Good quality music was not available on demand until the advent of the gramophone and the radio in the early twentieth century. By the 1950s, high quality recording studios and inexpensive playback machines made it possible to listen to some of the best musical performances on demand around the world.

From a musician's perspective, the need for elaborate recording studios and the cost of mass-producing musical records meant that very few performers could win a contract with a major label and become known to the public. The screening process did, on the surface, select the top talent, but in reality, the competition

> From Simorgh's perspective, entertainment is an overhead that is a cost of having human nodes.

was not open to all, and there was a significant built-in partiality.

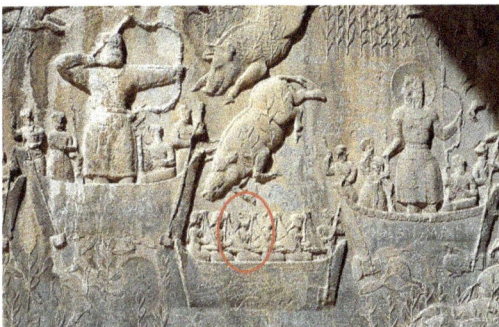

Figure 10-19. This bas relief depicts the Persian king's hunting ceremony. Musicians in the boat are shown with the strings of their harp-like instruments clearly resolved (circled). [131]

---

[130] "Archimedes", Encyclopedia Britanica.

[131] S. Parvin, B. Afkhami, "Sassanid Music (from Historical Texts to Archeological Evidence)," Tomsk State University Bulletin on Culture and art History, No. 37, 2020.

Recording, distribution, and playback of music have undergone radical simplification in the twenty-first century. High quality multitrack audio recording and mixing can be done with a relatively small budget, and the disbursement of music has become effortless on social media. This has led to the emergence of a plethora of new artists and subsequently has made the task of updating personal playlists quite challenging. Even professional music reviewers do not find a chance to sample more than a very limited number of music albums. Increasingly, automation and algorithms are selecting and recommending the playlist that the listener may like. The addition of AI will impress the customers by being highly on-target with selecting the right type of music according to the user's individual taste, though it might limit the ability of one's taste to evolve.

Despite automation and the loss of low-skill jobs, the entertainment industry is comparatively less prone to technology takeover, since entertainment is by and large a human to human interaction. The audiences enjoy both the talent and the imperfections of human artists, and relate to their life stories. From Simorgh's perspective, entertainment, like art in general, is an overhead that is a cost of having human nodes. We need to ensure that Simorgh's value system accepts and respects the existence of such overhead.

Law and Judgement

The search for legal precedents in court cases in the United States used to be a labor-intensive task that could occupy the time of dedicated teams of paralegals in law firms for weeks. This task is currently undergoing rapid automation.

We will have a fully automated and ever-present legal system, capable of detecting infringements and arriving at immediate judgments.

Many law firms relay on the services of a virtual "attorney" such as "Ross," powered in part by IBM's Watson artificial intelligence. Ross can function as a chatbot that conducts legal research on relevant laws and precedents and is designed to interact verbally with users such as fellow lawyers. It can understand verbal questions and provides specific conversational answers. Ross can perform the job of a team of paralegals at a fraction of the time.

Algorithms can be very useful in certain legal practices such as bankruptcy and divorce disputes, which usually require navigating lengthy statutes that have been interpreted in previous decisions. Algorithms will be able to analyze most scenarios and take into account exceptions, loopholes, and historical cases and recommend the best path forward. In the near future, many court cases will be prepared solely by AI algorithms, and human lawyers will only take over the highly specialized cases.

Eventually, settling civil disputes will not need lawyers or judges. Most cases will be filed and tried entirely online in virtual courts. In many such cases, the facts will be uncontested because digital traces of most activities and transactions are readily available. So, if you have a financial dispute with another person, each party would invoke a law suit app for presenting their case. By the time the information has been entered by both sides, the adjudication will be readily available, and the case will be closed on the spot. Even the execution of contracts, wills, and divorces will be automated, and the outcomes will be more likely to be fair and impartial than in current practice.

Further extrapolating this trend, it is likely that we will have a fully automated and ever-present legal system, with smart contracts and laws and real-time data, capable of detecting infringements and imposing immediate judgments. Such a system will become one of Simorgh's internal self-regulating mechanisms.

The criminal justice system has its own special requirements and relies heavily on human psychology and eyewitness testimonials that challenge the judgement process. To appreciate the advantages that AI can bring to criminal judgement, it is essential to keep in mind the severe shortcomings and biases of human judges and juries. Human judgement often relies on personal intuition and is heavily influenced by factors such as the person's mood and temperament. There are strict judges and lenient judges, and there are judges who may be susceptible to ideology, influence, and corruption. AI-based judgement promises to mitigate many such shortcomings. It is true that machine learning inherits the biases that are built into the fabric of the society, but at least it is less likely to add personal biases to the pool. Biases in algorithms may be detected by people and by other algorithms and may be corrected over time. The same cannot be said of biases in human minds.

Automated algorithms are already being used in the criminal justice system. An example is the application "Predpol" that helps with what is known as proactive policing. It optimizes police deployment based on crime patterns over the town. Another set of applications known as Northpointe's Risk Assessment Tools help the courts decide if and when to return offenders to the community without jeopardizing public safety. Pretrial jailing is a practice that decides if defendants will await trial at home or in jail. The judge must balance the cost and inconvenience of incarceration against the risk of flight or societal harm. It has been shown that pretrial detaining decisions can have significantly better outcomes when done by algorithms rather then judges. One study shows that crime can be reduced by close to twenty-five percent with no change in jailing rates if the decisions are made by AI algorithms.[132]

Should criminal justice be retributive, preventive, or rehabilitative? For many centuries our criminal justice system has been primarily retributive. This system contains elements of vengeance from our savage past, but it also removes threats from society, albeit after some damage has been done. In addition to the removal of the threat, retributive justice also serves as a deterrent to future crime.

The motion picture *Minority Report* (20th Century Fox, 2002) depicts a futuristic version of preventive justice where people are arrested based on their intentions to commit a crime. This is an attempt to

---

[132] Jon Kleinberg, et. al., "Human Decisions and Machine Predictions", National Bureau of Economic Research, February 2017.

remove the threat from the society before any damage is done, but its implementation in the movie is extreme and obviously unacceptable. However, it is possible in some cases to predict a high likelihood of crime and change the circumstances that may lead to the crime. Preventing the possession of fire arms by emotionally unstable people is an example of preventive justice. Foiling a terrorism plot is another example. The availability of public data on people and their intentions in the future will make it easier to intervene in specific cases to prevent a crime. Preventive justice will become more prevalent in the society as better technology enables it.

Rehabilitative justice, which is being practiced to some degree today, treats crime as a social disease and attempts to "cure" criminals and remove their proclivity to committing crime. This is a humane approach to justice, but it is not an effective deterrent. It appears that retributive, preventive, and rehabilitative justice will continue to be practiced together in the near future, though automation and the availability of information will shift the emphasis away from retributive justice to the other two approaches.

Currently there are two levels of judgement to be made in most criminal cases. The first is to positively identify the guilty party. The second is to give the appropriate sentence to the offender. Identifying the guilty party used to be a cumbersome and lengthy process that included searching for a motive and listening to lengthy testimonials. Recently, the data trails and the digital footprints that people leave behind, are transforming the court trial into a data search that is best done by AI than people. Furthermore, examples given above and studies, such as the pretrial jailing research, strongly suggest that deciding the appropriate sentence is also much better done algorithmically and by AI than by people.

Employment in the field of legal services is still on the rise due to population growth and the diversification of business. In the long run however, employment in this field will inevitably shrink due to automation. The provision of advisory legal services to businesses is expected to continue for a longer period of time.

Corporate Management

Corporate management in various fields is a position that is less prone to automation in the near term and will continue to be held by actual people for some time. Let's investigate the longer term future of corporate management in view of advancements in technology.

Some of the major responsibilities of a corporate CEO are the following:

- Maintain or bring the company to a healthy level of profitability and growth.

- Build or maintain a favorable public image for the company.

- Be an inspiration to the company's employees, and develop an overall company vision.

- Make high-level decisions about policy and strategy,

- Build alliances, partnerships, mergers, and acquisitions with other organizations.

- Recruit qualified upper management staff to run major operations of the company.

- Be the primary spokesperson for the company.

- Oversee the company's fiscal activity, including budgeting, reporting, and auditing.

- Report to the board of directors and the investment community.

- Assure compliance with laws and regulations.

- Identify and address problems with company operations.

- Identify and take advantage of business opportunities for the company.

Many of the responsibilities listed above revolve around communication skills and effective interaction with people. The assumption is that other companies and the investment community are led by humans. If they were AI-driven instead, most of these functions and responsibilities of corporate management would not exist. An example is public speaking to promote the image of the company. This would not be necessary if facts were communicated by direct data exchange among companies and software algorithms analyzed the data. But before all management becomes AI-based, the pioneering companies still need to have a way to deliver AI-generated speeches to humans who oversee other organizations, or to those who hold key positions within the company.

Inspirational public speaking has long been considered a form of art. However, increasingly speakers are being assisted by opinion polls that reveal what the audience cares most about. Writing aides, known as sentiment text analysis software, can decide what is considered inspirational by most people.[133] Currently AI-written and delivered speeches tend to be only objects of curiosity. Typically, the audience would find it very disappointing if they thought a human was giving the speech. However, rapid and steady progress is being made in this field, and soon, this deficiency of AI will be alleviated. Impressive inspirational speeches will be given by company avatars instead of human CEOs in a very convincing manner.

Despite such advances in AI, corporate governance is likely to remain human centric for some time in the future. Computer-aided management will perform many of the background tasks to make the job of corporate management much easier for people in those positions.

Education

Both in the wild and in human societies education used to be a part of children's upbringing and it involved teaching them the knowledge and skills they needed to live an independent adult life. Parents and close relatives were the teachers. With specialization came the concept of apprenticeship, which provided career options

---

[133] See for example the speech writing assistance provided by intentex.com.

beyond one's family occupation. Literacy was also initially taught through apprenticeship for some government functions. Later, in the West, religious organizations came to dominate literacy instruction, which was required for reading sacred scriptures and became a component of general religious education. Other subjects such as history, literature, and philosophy were added to expend the scope of the religious curriculum. With the spread of secularization in recent centuries, governments recognized the importance of literacy for societal cohesion and started to establish non-religious schools with a standard curriculum.

The current system of age-based classroom instruction, that evolved from the so-called Prussian system, has been in use successfully for a couple of centuries with minor variations around the world. Motivated students and those fortunate enough to have good teachers benefitted more from the system than others.

Lately, we have seen that learning a subject through online instruction is more effective for most students than in class learning. An online class can be taught by the nation's best instructors, and has the added benefit of being self-paced. Local teachers can then focus on helping each student individually with both academic and social aspects of learning. This approach is being promoted by organizations such as Khan Academy and Coursera. Technologies such as VR and AR will further enhance online learning by offering an immersive experience.[134] It is not surprising that VR can enhance the retention of information to as much as seventy-five percent compared with only five percent from a standard lecture.[135]

From an employment perspective, we can expect a decrease in the number of lecturer positions in favor of educational content development jobs for designing and creating sophisticated remote instruction packages. But overall teaching-and counseling-related

---

[134] Y. Slavova and M. Mu, "A Comparative Study of the Learning Outcomes and Experience of VR in Education," 2018 IEEE Conference on Virtual Reality and 3D User Interfaces (VR), 2018.

[135] VR Learn: Virtual Reality & Learning, by the National Training Laboratory 2017.

jobs are expected to persist for a long time due to people's needs for socialization and human contact as a part of their education.

Higher education and specialization are also changing rapidly. The concept of lifelong learning is hardly new in education, but it is becoming more relevant as people's knowledge becomes obsolete at a faster pace every day. The value of an earned university degree diminishes rapidly with time, while job experience, micro credentials, and professional certificates become more relevant later in one's career.

People who will ultimately become nodes of Simorgh need to initially have enough expertise to be employed at a technology-driven organization that will eventually transition to Simorgh. The qualifications required for employment will vary depending on the nature and multiplicity of Simorghs, but it is safe to assume that a minimum level of education and specialization will be required. Education and training will remain important functions within Simorgh. Learning by human nodes of a fully-fledged Simorgh will be lifelong and continuous. Human nodes of Simorgh just like non-human nodes will need to keep improving and adapting. The quest for improvement has been present throughout human civilization. However, noticeable improvements that used to take generations, now need to happen on a day-to-day basis.

Projection

A report titled "The Future of Jobs" by The World Economic Forum in 2016 argues that the fourth industrial revolution marked by advances in Automation, AI, robotics, biotechnology, etc., will precipitate significant job losses throughout the economy. Table 10-2 summarizes some of their predictions. Others make similar projections, but on a less aggressive timetable.

Occupation	Loss Automation
Warehouse and inventory workers	98%
Legal Assistants and paralegals	98%
Accountants and bookkeepers	97%
Local government administrative workers	97%
Salespersons and retail assistants	95%
Utility company engineers	95%
Pharmacists	94%
Store cashiers	90%
Couriers and postal workers	86%
Taxi drivers	57%
Interpreters and translators	32%

Table 10-2. Percent job losses expected in each category within two decades. (World Economic Forum 2016).

Future of Employment and Transition to Simorgh

Let's consider the following historical trends:

- Much of the household chores and farm work used to be done with the help of servants and slaves in the not too distant past. This practice has gradually ceased, and in most countries, slavery has been outlawed.

- Up until about 150 years ago, the majority of jobs in the United States were in agriculture, before machines like tractors, combines, and fruit-picking equipment automated the majority of such jobs and displaced a large portion of the paid farm workers. (Figure 10-20).

- Trains and automobiles initially augmented the means of transportation that were predominantly horses and carriages at the time. Eventually, automobiles eliminated the need for horses altogether.

In this chapter, we have examined how automation is causing most existing jobs to be performed with far fewer workers.

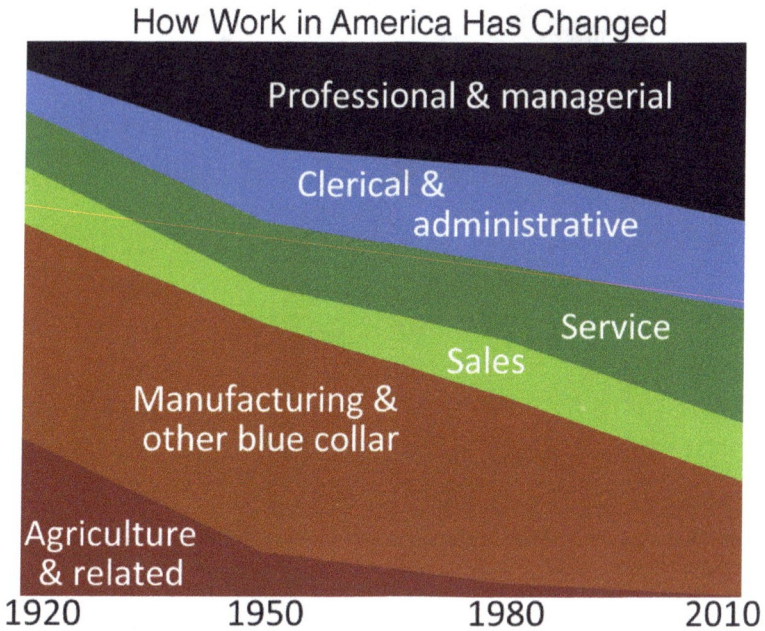

Figure 10-20. Occupational changes in the US over a 90-year period. Most notably jobs related to agriculture have been decimated and manufacturing jobs have contracted (Statchat, University of Virginia April 6, 2012).

The weakening demand for human labor will initially result in the lowering of weekly work hours. This may be seen as a positive outcome. The 32-hour work week is likely to be adopted soon in industrialized countries. However, increasingly in many technology-based industries, the concept of official work hours is losing its relevance to the specialist staff. A day's work is never done for such employees, and many companies offer them flexible work hours. The concept of working hours is becoming less well-defined as an increasing number of projects require international teams to collaborate in different time zones. It can be envisaged that work and personal business will be mixed at any time of the day for Simorgh's nodes, while most unincorporated humans will continue to follow stricter official guidelines of working days and working hours.

Another tool to make underemployment more tolerable is what is known as the Universal Basic Income (UBI) that may be adopted in some countries. This is an inverse tax through which everyone receives a nominal monthly salary from the government. This may be necessary for some governments in the short run to prevent any potential uprising by the unemployed population. UBI may be sustainable if the government's budget has a significant annual surplus or if there is a single world government that doesn't face international competition. Otherwise, Countries that implement UBI will find it to be an undue burden on their annual budgets and competitiveness. This is because the direct return on UBI payments in the form of increased production is very low. It only maintains a minimum level of consumption, which will have economic significance, but with little positive impact on industrial growth. UBI is likely to increase productivity in the non-tangible goods and services with esthetic, artistic, and humanitarian value that people with free time tend to engage in. This may be found appealing to the local population, but it will be weak in an international competitive environment.

Countries with smaller populations that are highly trained and productive will not be burdened by UBI and will have a distinct budgetary and competitive advantage. The inevitable consequence is that there will be a significant pressure to reduce population growth. Incentives and penalties will be implemented to encourage smaller families or to postpone reproduction until later in life. It is thus natural to see a significant decline in human population as automation continues to grow. It is noteworthy that the technology whose purpose is to make life easier for the human population will eventually render a significant portion of the population superfluous, though it will make life easier for the rest.

The well-trained human population will have full-time employment in organizations in which decisions are made collectively by people and algorithms, and not by individuals. Human figureheads may still retain key executive positions, but their role will only be to give a stamp of approval to decisions already taken by established procedures. Every employee will be highly specialized and will not necessarily see how the organization may benefit from the outcome of his work.

When a large organization - which may be a company, a group of companies, a country, etc. - forms comprehensive strategies not championed by any particular person or committee, the organization starts to assume a Simorgh identity. When no one supervises the development of plans, strategies still continue to emerge, but then it is really Simorgh who is making the strategic decisions. People employed by these organizations will form the human nodes of the organization, or the human nodes of Simorgh. They will in essence be Simorgh's body cells, and each individual cell will not be aware of the intentions of the whole body. Simorgh will have human nodes and nonhuman nodes. The nonhuman nodes will be both stationary and mobile computers.

> When no one supervises the development of plans, strategies continue to emerge, but then it is really Simorgh who is making the strategic decisions.

Life as a Human Node of Simorgh

If you are a node in Simorgh's body, your workplace responsibilities are assigned to you by algorithms that specify the goals you need to achieve, and provide the work detail that you need. You receive information from various sensors and from other nodes who are your coworkers and work on similar projects with similar goals. You know how to process such information according to your expertise and with the tools at your disposal. You disseminate new information and/or direct actions as needed. Your attention is focused on your projects and you know the end goals to be achieved. But you may not be aware of Simorgh's presence and the reason for its interest in your work.

You don't have to commute to work – where you live is your workplace. There are no work hours. Work is on an as-needed basis. You are always connected to the network, and you are called to action when needed. This is reminiscent of old lifestyles associated with livestock farming when farmers had to take care of whatever was needed on the farm whenever it was needed. The notable difference is that you do not need to engage in any significant physical activity. You instruct robots and automated equipment to do so. With virtual reality, your mobility is optional. You may send your avatar robot to examine places where sensors may not already be in place. Your salary is a number that is transferred to your "smart bank account." It orders everything you need for you, and the ordered items are delivered to you. You can order discretionary items or entertainment by yourself whenever you need spontaneity.

> People need to keep pace with technology advancement. Specialization is what is needed initially, but eventually a deeper structural adaptation will be required.

Your health and security are automatically taken care of. When Simorgh is not in crisis, you have plenty of free time to play and enjoy. This is the simple picture of life as a human node in Simorgh's body. This is an extrapolation of today's corporate employee lifestyle, and this is how the transition to Simorgh will start, but it will not stay this way for long. Human nodes of Simorgh need to keep pace with technology advancement. Specialization is what is needed initially, but eventually a deeper structural adaptation will be required.

Robots and artificial intelligence have been evolving to be of help to humans. In the future, both robots and humans need to evolve together to assure Simorgh's prosperity. Humans need to evolve by gene manipulation and acquire enhanced capabilities by the incorporation of mechatronic body parts. In the absence of such evolution and improvement, human nodes will not be able to keep up with the accelerating pace of technology development. Simorgh's human nodes need to become "GM hybrids" who are in fact genetically modified cyborgs (Figure 10-21).[136]

The untrained or low-skilled people will not have full-time employment and will not be bound by the algorithms that run Simorgh's nodes. Thus, their function may vary significantly from day to day. This group constitutes the unincorporated population

---

[136] A cyborg or "cybernetic organism" is part human, part machine.

that does not merge into Simorgh. Technology will split the human society into Simorgh's nodes and the unincorporated.

In a short period of time, Simorgh's human nodes will become highly differentiated, in both appearance and function, from the unincorporated humans who will continue to thrive on a lower evolutionary scale. Many humans may prefer the unincorporated lifestyle, and many may have no choice but to adopt it.

It should be emphasized again that neither the human nodes nor the unincorporated humans will have the ability to comprehend the nature and intentions of Simorgh once it has fully evolved.[137] The inverse is also true that Simorgh doesn't need to pay much attention to humans or human groups outside of its structure unless they attempt to cause significant harm.

Figure 10-21. In the old TV series *Bionic Woman*, the character Jaime was given acute hearing and extraordinary arm and leg strength through "cybernetic" implants.[138]

---

[137] Even in current human societies, citizens of each nation remain in the dark about the true role that their country plays on the international scene. Each government presents a favorable view of itself to its own citizens. Most people think of their country as a force for good in the world, while most international deals and informal agreements remain hidden from the public.

[138] Rickey Price, "Please Stand by: A Journey through the History of Television Super Heroes," Comic Watch, November 23, 2019.

Summary

Industrial automation started in the areas of manufacturing and agriculture, and even though it displaced many workers from farms and factory floors, the extra productivity created additional jobs in distribution and other services. The net job creation was positive. Since the end of the twentieth Century, automation has begun to infiltrate the service sector. Job losses in the service sector do not lead to sufficient job creation elsewhere, and thus cause a net loss in low-skilled jobs. The extra demand created by lower automated service prices generates additional revenue for the service owners, but not much additional employment.

Most traditional jobs performed by low skilled workers will not survive rapid advances in technology. One exception is the class of jobs such as arts and entertainment that appeal to human emotions. Such jobs will continue to flourish among the segment of human population that does not incorporate into Simorgh's body (unincorporated humans). Technically well-trained people and specialists will gain permanent employment and may become nodes or body cells of Simorgh. Eventually, they will differentiate both physically and functionally from ordinary lower-skilled unincorporated humans.

Chapter 11: Specialization, Equality, and Personal Autonomy

Professional specialization has been one of the hallmarks of human civilization since its early days. Many distinct classes of city dwellers emerged based on their occupation. They included grocers, soldiers, bakers, and craftsmen, to name a few. Over time, city inhabitants continued to refine their fields of expertise, and the number of such fields grew. The emergence of industrialization and the development of new technologies accelerated the creation of more differentiated areas of specialization. A clear example is a modern hospital, where hundreds of specializations have replaced the old single occupation of a "medicine man." Training is not the only factor that leads to specialization. Inherited genes and one's upbringing jointly create a predisposition for attaining certain skills. For example, not everyone is naturally able to become a good musician or athlete.

One specialization that predates civilization and is solely determined by one's genetic makeup is childbearing. The same predetermined roles for reproduction exist in most living species due to the need for genomic diversity in the offspring. In some species, the deterministic roles go beyond reproduction and extend to other responsibilities. For example, worker ants and soldier ants perform distinctly different tasks predetermined at birth. Their genetic composition also determines their distinct appearance. This "eusociality" order can be found in a variety of insects, crustaceans, and mammals. (Figure 11-1).

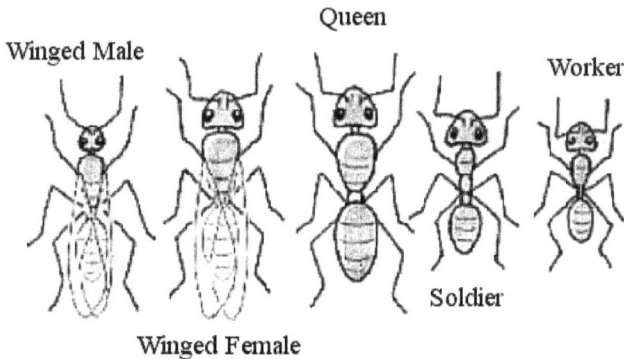

Figure 11-1. Eusociality in ants: different social functions are performed by ants that have distinctly different appearances.

In earlier human societies, occupations tended to stay in families, and the necessary skills were passed down to the children through private apprenticeship. This led to the formation of closed and rigid social stratifications based mostly on occupation. The rigidity was enhanced through social customs of heredity and endogamy, which is the practice of marrying within one's social class.

One of the more unyielding examples of social stratification existed in India and is known as the "caste" system (Figure 11-2a). Similar hierarchies emerged and still persist to some degree in Europe, U.S., and other parts of the world (Figure 11-2b, c).

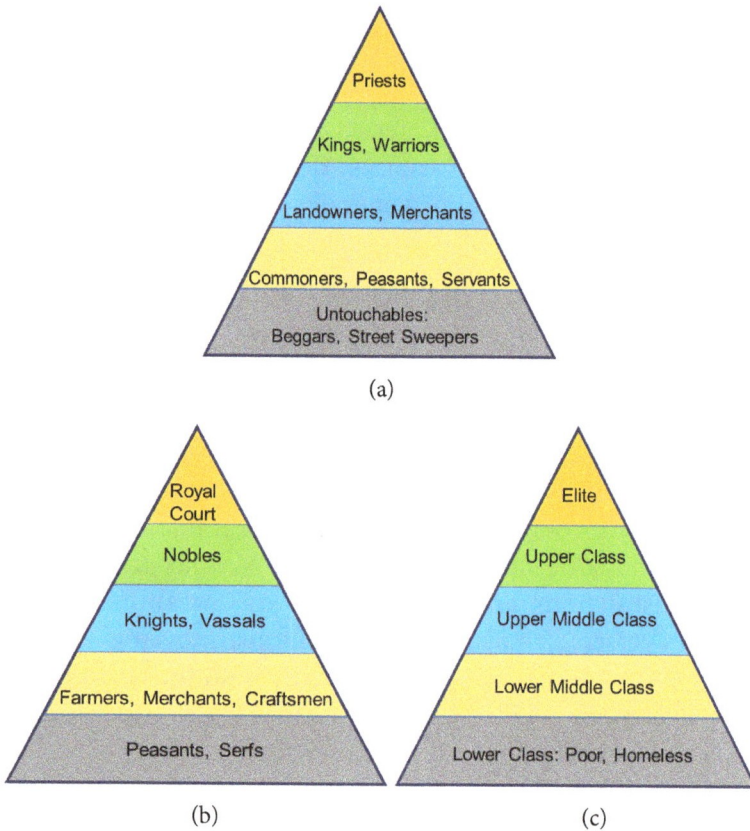

Figure 11-2. Class hierarchies in various parts of the world: (a) Caste system in India, (b) Traditional class system in Europe, (c) Less formal class structure in the U.S. and many other countries. Modern societies encourage mobility among classes and strive to prevent the formation of "closed" classes.

The vastly different living standards of the various social classes is viewed by many as an injustice that needs to be corrected. Egalitarianism is a quest to remedy social injustice and to administer policies to minimize it. Egalitarianism comes in many flavors: philosophical, social, economic, gender, and luck egalitarianism, to name a few.

The current widespread desire for equality favors establishing only equal opportunity and assigning equal worth to all people, but not necessarily a desire for economic equality. The inclusion of these goals in government policies is a relatively recent notion that has gained increasing popularity since the eighteenth century. Prior to that, egalitarianism was primarily philosophical and preached by some religions. No practical steps were taken for its promotion, except by encouraging charitable donations. Additionally, the equality preached by religions generally did not encompass gender equality.

The concepts of economic equality and the formation of a classless society have been the subject of multiple social experiments throughout history. For example, there was a religious sect of Zoroastrianism that emerged in the sixth century AD, led by a priest named Mazdak.[139] It posited that God provided the means of subsistence on Earth for all humans to share equally, but the strong had managed to coerce the weak, seeking domination and causing inequality and suffering. The followers of Mazdak tried to abolish all class distinctions and share everybody's wealth. This sect is sometimes referred to as a proto-communist movement that tried to abolish all social classes. The spread of communism in the early-twentieth century was the largest sociopolitical experimentation of a classless society, involving a significant portion of the human population, albeit in underdeveloped countries. Remnants of communism still exist in some regions of the world, but mostly in title only. In practice, there is much more support for socialism concepts and for benevolent governments than for the communism ideal of a classless society.

Technological specialization is working against social equality. It is a force that is splintering human society into two major classes: the high-skilled minority and the low-skilled majority. Increasingly, the low-skilled majority find it more difficult to gain permanent employment in technology-dominated, fast-growing industries. On the other hand, the high-skilled minority is becoming further specialized and relies more heavily on automation and technology-based assistance in all aspects of life.

---

[139] Frye, R. N., "Chapter 4", The political history of Iran under the Sasanians, The Cambridge History of Iran, 3, Cambridge University Press, (1983).

The accelerating rate of professional specialization in areas of rapid technological development shows no signs of abating. A high degree of specialization causes the specialists to become preoccupied with narrow tasks in their daily routines. This naturally limits their awareness of the broader results of their work. The designers of integrated circuit chips, for example, do not necessarily know much about the computer programming that their work enables. This is a clue to Simorgh's development, as individual cells do not need to be aware of the overall activities of the body.

Specialization is poised to take another quantum leap forward, afforded by the emerging technologies of DNA manipulation and gene editing.[140] Future specialized humans with enhanced genes will have vastly different capabilities and appearances. Human soldiers and human scientists will not only function differently, but also have distinct appearances, reminiscent of the differences between human males and females. CRISPR/Cas9 gene editing technology has opened the door to the eventual polymorphism and differentiation of humans into specialized body nodes of Simorgh. Highly trained and specialized people will initially be Simorgh's undifferentiated cells (nodes) who may fit into different occupations of their choice. This will be the case during Simorgh's early stages of development. Early human nodes of Simorgh will be similar to stem cells, with the potential to fit into a variety of body functions. As Simorgh matures, the stem cells need to differentiate. The differentiation process will start when genetically modified humans emerge with traits customized for their specialization. Switching to alternate specializations will become difficult and differentiation will become permanent. This will inevitably bifurcate humanity. The majority of people will not follow this trend and will continue to live as unincorporated humans inside or outside of Simorgh.

---

[140] For more on gene editing, refer to Chapter 8.

Each person's field of specialization will determine what body function of Simorgh he or she will join. Genetic manipulation will enhance this specialization into a permanent genetic differentiation. Humans are already divided into men and women, who are genetically differentiated based on their reproductive roles. There are additional social and occupational differentiations based on inherited abilities, such as physical strength, intelligence level, and so on. Genetic modification will make possible the enhancement of current differences in physical traits. Large people may be made larger, small people smaller, and intelligent people given a higher intelligence level that will continue to improve. People with enhanced eyesight, hearing, or dexterity may become distinguishable from a distance. The ethical concept of social equality will be of no interest to Simorgh. Egalitarianism will inevitably fade away. Genetic differentiation will lead to closed social classes that will be accepted by the population, just as people in the past accepted and adapted to the presence of social classes like the caste system. One significant difference is that all of the new types of humans will be economically secure and well cared for as nodes in Simorgh's body.

> Highly trained and specialized people will initially be Simorgh's undifferentiated cells. Differentiation follows with genetic modification.

Freedom

Personal freedom and autonomy are other major topics of discussion within our society. Most people are familiar with the term "freedom" and generally think of it as the absence of coercive constraints. But a precise definition of freedom is less obvious because it pertains to so many diverse facets of life. There is political freedom that is a favorite topic in news media; there is economic freedom that comes with wealth; there are civil liberties and civil rights that need to be defended. Much focus is placed on the role of the government in either delivering or protecting such freedoms.[141]

What is known as personal autonomy is the ability to plan and execute one's short-term and long-term schedules practically but without constraints. It is a broad concept and assumes the presence of all the listed freedoms. In today's society, personal autonomy is more directly influenced by workplace policies and each person's accumulated wealth than by government rules.

To further expand on the difference between personal autonomy and personal freedom, it is worth noting that freedom is typically referred to as people's ability to engage in a particular activity of their choosing. Examples are freedom of expression and freedom of religion. Personal autonomy, on the other hand, refers to self-determination and not being restricted in one's daily life. People who have gathered enough wealth to achieve "financial independence" for their desired lifestyle enjoy the autonomy of deciding their own daily activities and setting their own schedules. Those who rely on their employment to cover their everyday expenses are likely to enjoy far less self-determination. They may exercise some limited autonomy during their leisure time to the extent permitted by their financial wherewithal.

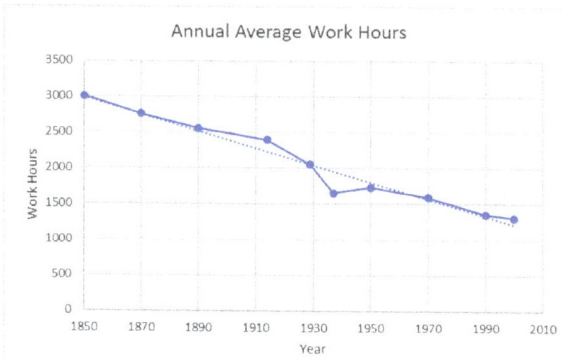

Figure 11-3. Average annual employment work hours in the U.S. since 1850. The fitted trend-line shows a reduction of approximately 11 hours per year. (Angus Maddison, "The World Economy: A Millennial Perspective," OECD, Paris 2001).

---

[141] One person's freedom may infringe on the freedom of another. The legal framework that ensures people's freedom without conflict is known as liberty.

One measure of leisure time is the number of required work hours. The actual number of work hours per year has been on a steady decline for many years (Figure 11-3). Also, in modern societies, there is a limit on the time commitment per week that an employer can demand from each employee. The forty-hour work week is a guideline that provides a degree of autonomy in every employee's life. However, many technology-based companies are currently not strictly adhering to this guideline, particularly in jobs that require a high degree of specialization and creativity. Employees in these companies are expected to be "accessible" to address problems or engage in teleconferencing at any time of the day. This trend is likely to continue, blurring the border between work hours and leisure time. This phenomenon, together with the fact that automation is reducing the actual workload on humans, provides a glimpse into the future lifestyle of a human node in Simorgh's body: the node's connection to work will be continuous, but the workload will be light because of automation. There will be some low-stress daily assignments mixed in with on-demand work at any time of day. There is a striking parallelism with the work style of the average cell in the human body. On average, the workload of each cell is moderate except in times of crisis.[142] Overall, the future human nodes of Simorgh will be well cared for. They will have a light workload and plenty of leisure time, but they will not have much freedom to change the course of their lives.

> The ethical concept of social equality will be of no interest to Simorgh. Egalitarianism will inevitably fade away.

Summary

Egalitarianism as an ideal may continue to survive in societies outside of Simorgh's body. However, in the body of Simorgh, there will be genetic diversity and no equality. There will be comfort and leisure time, but there will be no freedom to switch careers.

---

[142] Of course, there is considerable variation in the workloads of various cells. For example, the work schedule of a heart muscle cell is much different from that of an earlobe cell, even though they receive the same overall benefits.

Chapter 12: Simorgh: Its Identity, Control, and Thought Patterns

The first sign of Simorgh's emergence is the gradual removal of individuals from institutional decision-making. This is already underway, and it manifests as the streamlining of decision making through digital assistance. Let's explore the mechanisms of decision making more closely.

Decision-making process

Decision making is the act of choosing among alternative courses of action or alternative answers to a question. Humans have always used a combination of analytical and emotional tools when making both routine and critical decisions. The analytical tools include the use of historical data, current data, and future projections. Current data is the real-time observation of the situation that requires decision making. Historical data is the memory of similar situations in the past. Future projections and simulations are mental images that depict potential outcomes and consequences. For example, when someone spots the threat of an attack in the wild, the recognition of the predator and the judgement of its posture and distance are parts of the current data. The decision to be made is either to hide, attack, or escape, with variations such as distraction, etc. Experience or observation of similar encounters forms the historical data, and the planning of the attack or the visualization of the escape route provides the future projection or simulation. Fear and intuition are the feelings that accelerate the decision-making process. Intuition, or "gut feeling," is a feeling that is shaped by past experiences and reflects a person's adopted approaches and preferences. Though not infallible, intuition allows one to evaluate the options much more quickly (Figure 12-1).

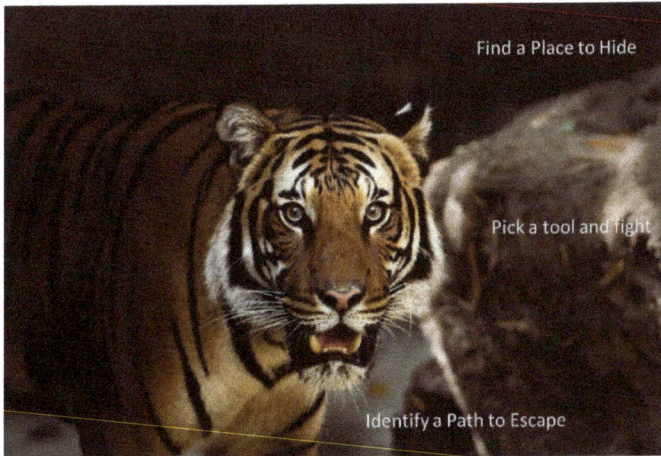

Figure 12-1. The need for quick decision-making when facing danger. Fear and intuition shorten the reaction time.

Of course, not every decision is as urgent or involves such life-threatening circumstances. Most decisions are routine and repetitive such as when to turn on the dishwasher or what grocery items to buy. Routine decisions within a company, such as when to schedule a business trip, are referred to as operational decisions. Simple mental or computer algorithms can normally make operational decisions efficiently. There are rare exceptions, such as the travelling salesman problem, whose objective is to minimize overall travel distance when visiting multiple cities. A simple algorithm for arriving at the correct sequence could for example be "Visit the next closest city that you haven't visited." It doesn't take many examples to see that this rule doesn't work, and to date, neither does any other simple rule. When the number of cities is large you don't want the brute force method of calculating the distance for all the possibilities and picking the shortest one; but since there is no efficient method for solving this problem, you plan your travel without regard for this optimization. In other words, you don't include this goal into your decision making process.

There are cases where decision making affects many people. Top government officials and company executives routinely face such instances of making strategic or tactical decisions. Traditionally, their assistants and subordinates collect the relevant past and present data, provide future projections, and summarize the decision options. When this information reaches the decision maker's desk, biases and preferences of the preparers are already imbedded in the documents, and the choices have already been narrowed down. The final decision maker usually adds his own additional emotions, intuitions, and biases to the process. Recently, much of the work of the assistants has been automated, and decisions can be made more quickly thanks to electronic data processing and management algorithms. Instead of waiting for input from assistants, the final decision maker can have all the information he needs at his fingertips with much less processing time.[143] More recently, automation has been able to go one step further and identify and itemize the decision options as well as the tradeoffs for each. The human decision maker then simply selects from the listed options. In the near future, with the automation of the

> Decisions made without guidance by any individual, signal the dawn of the age of Simorgh.

selection process itself, the role of the human decision maker will first be reduced to a figurehead for approval, and later it is likely to be eliminated altogether. This will gradually remove most human emotions and biases from the process, which could be a positive outcome.

---

[143] M.J. Kochenderfer, T.A. Wheeler, and K.H. Wray, *Algorithms for Decision Making*, MIT Press, 2022.

Automated decision making is currently being practiced, though perhaps not noticed by many. For example, in trading company stocks, what's known as "automated sentiment analysis" gathers and analyzes positive and negative public opinion about companies from the news and various web sources. Later, this information and other automatically collected market data are used by trading algorithms that make automated decisions on buying and selling company shares without human involvement. Stock market trading is highly time sensitive, and trading companies invest heavily in high-speed data networks to gain a few milliseconds of timing advantage that allows them to place an order ahead of a competitor. Now, considering the fact that a person takes over 200 milliseconds to just press a button, it is not surprising that humans need to stay out of the process.

A more subtle example is the path of technological development, to which many people contribute but no one is in charge and the path is unpredictable. Long term projections of technological progress are often proven wrong. Another example with procedures that can be examined is the gradual improvement paths taken by large industrial and weapon systems. For example, the military purchases armament with government approval. The weapons are tested either in simulations, military exercises, or in actual battlefield conditions in minor armed conflicts. Algorithms analyze the exercise and identify the relative importance of the various military systems in the battlefield. They also identify the weaknesses of each system and provide feedback to the manufacturers. In turn, the manufacturers offer an upgrade path based on the findings, and the plan is approved. The direction of the development and the military's future capabilities are therefore algorithmic reactions to tests and exercises. It is a gradual evolutionary development and it is not planned in advance by any single person, even though many people contribute to its planning, and signatures by people in command suggest thorough examination at each stage of approval (Figure 12-2). In fact, no one is compelled to challenge either the procedure or the recommendations.

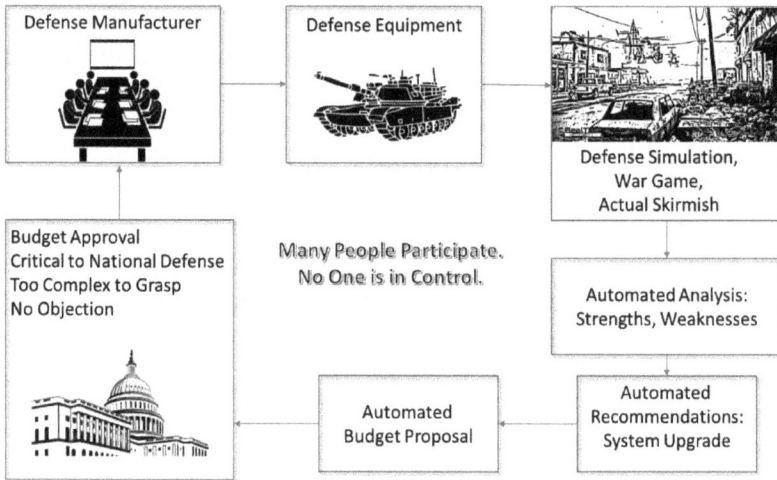

Figure 12-2. A hypothetical process of decision making in military equipment and systems development. Though many people participate in the process, the cycle runs by itself and in a sense, has a "mind of its own."

In companies and various operations in our society, we are beginning to see similar phenomena, where decisions are made collectively without having been planned or directed by any single person. Specialists make micro decisions along the way, but the macro decision making process is spontaneous and unplanned. In other words, people are increasingly involved in only the fine details of projects whose overall directions are algorithmically driven. This phenomenon signals the dawn of the age of Simorgh: Decisions are made extemporaneously without guidance by any individual. We are observing the emergence of an early Simorgh who is learning to make decisions on its own. This is analogous to decisions made by the human brain, where many brain cells are involved but none are in charge or aware of the decision being made.

In our transition to Simorgh, each human becomes a cell, or node, in the body of the larger organism that then makes all the strategic decisions independent of each participant. This relieves each individual of the burden of making broad choices and allows them to focus on their area of expertise. This evolutionary transition is analogous to older examples in which individual elements merged to form a larger entity, such as the joining of individual cells to make multicellular organisms. A more complex and capable being emerged in each case and Simorgh will be no exception.

Analogy with Living Organisms

Consecutive evolutionary stages of life on Earth have been occurring at an accelerating pace. As shown in the table below, it took billions of years for early transformations to take place. Later stages took millions of years. Transformation from civilization to Simorgh is only taking thousands of years, the latest phases of which are being further accelerated by major technological advances such as networking, gene editing, and artificial intelligence.

Prokaryote	⟶ Eukaryote	(2,000 million years)
Single Cell	⟶ Multicellular	(1,000 million years)
Multicelluar	⟶ Animal	(1,000 million years)
Animal	⟶ Human	(600 million years)
Human	⟶ Civilization	(200 million years)
Civilization	⟶ Simorgh	(0.01 million years)

Table 12-1. The rate of evolutionary transformations on Earth is accelerating.

The first step towards the formation of Simorgh was the birth of human civilization and the emergence of urban living several thousand years ago. Cities and countries created organizations that function, in a rudimentary way, like human body parts (Table 12-2). There is a "brain" in the form of the government that makes major decisions. There are defense systems, food and water distribution networks, waste removal mechanisms, and so on. These sectors of the society work together as if they were loosely bound body parts of a larger organism. So far, information sharing within and among these body parts has been slow and limited. It is only recently that an interconnected "nervous system" has emerged in the form of the internet. Through the internet, all compartments and members of the society are becoming more tightly connected to each other. In fact, the formation of this nervous system is acting as the needed "thunderbolt" that will bring Simorgh to life.

Human Body Part	Urban Analog
Brain	Centralized Government
Blood Vessels	Roads, Waterlines, Electrical Lines
Skin	Borders
Immune System	Defense System
Endocrine System	Broadcast Announcements, TV, Radio
Nervous System	Mail, Telephone, Internet

Table 12-2. Some human body functions and their analog in an urban society.

Not only do city and country branches function similarly to body parts; each government bureau and large private organization has its own departments and divisions that serve the entire unit in the same way that body parts do. It is not clear which segment of the human society will be the first to assume a Simorgh identity. The smallest entity that has the required subdivisions is a large organization or a sizable corporation. The next larger candidates are cities, countries, or perhaps a bloc or group of countries with a shared communication network. Finally, there is the remote chance that the whole world would emerge as a single Simorgh.

Compared with a country, smaller entities such as a large corporation have an efficiency advantage in their transition to Simorgh: A corporation can be selective with its choice of personnel, and it can limit its costs by not hiring low productivity employees. This will allow the entity to evolve rapidly and efficiently. Countries, on the other hand, do not have this luxury, at least not in the beginning. A low unemployment rate is expected to apply to a country's entire population. Due to the overhead of supporting low productivity workers, a government may experience a slower rate of progress in transitioning to a Simorgh than a corporation. Simorgh doesn't need the non-specialized population and has two options for dealing with them: One is to offer them full employment at the expense of efficiency. Non-specialized people who occupy permanent positions within Simorgh will act as low-productivity nodes of the body, similar to excess fatty cells in the human body, and will lower the efficiency and progress rate of Simorgh as a whole. The other option is for Simorgh not to offer employment to low-skilled populations. Eventually, a separate economy will be formed by non-specialized workers whose employment will primarily support their own needs. They can still interact with Simorgh in a capacity similar to the interaction of microorganisms with the human body.

Depending on its size and early programming, a Simorgh may have a central command-and-control operation, or it may not. It is worth remembering that there are successful living organisms such as the jellyfish (Figure 1-4) that function on reflexes alone without any central control or brain. Similarly, it is conceivable that a Simorgh may be able to function on reflex-type algorithms without any central decision-making unit. A large company may evolve into this type of a Simorgh if its products have a stable market and no strategic central planning is needed. Examples are rare, but a large international transportation company may be a candidate. Most likely, other entities will maintain a central command and control operation that allows them to do strategic planning. It is worth emphasizing that even in that case, decisions will be made by algorithms with input from many human and non-human nodes. No single person or single node will be in control of any institutional decision making.

Danger to Humans?

There is a long-term risk that Simorgh will expel all of its human nodes and rely solely on AI-driven computers and machines. There is also the danger that Simorgh might perceive unincorporated humans as threats and try to eradicate them. In recent decades, many authors have expressed concern about autonomous machines in general and their potential to enslave or eradicate humanity (Figure 12-3). Isaac Asimov, a prolific science fiction writer, wrote a book in the mid-twentieth century titled *I Robot,* which focused on the possibility of encountering rebellious robots in the future. He devised the following three laws for programming robots to prevent such a prospect from materializing:[144]

---

[144] The adopted laws ultimately failed even in his own book.

1. A robot may not injure a human being or allow a human being to come to harm through inaction.

2. A robot must obey the orders given by human beings, except where such orders would conflict with the First Law.

3. A robot must protect its own existence, as long as such protection does not conflict with the First or Second Laws.

Of course, these laws require substantial revision in order to become applicable to today's AI programming. In their original form they may look simple, but their implementation is not as easy as it may appear. For example, simple words such as *harm* and *human* are challenging to define for a machine. Does *harm* mean only physical, or does it include emotional harm as well? Is a dead or incapacitated human still considered a human? Should the robot obey orders given by *any* human regardless of credentials and age? Asimov's pioneering laws leave out a lot of detail, but they draw attention to the need for such a framework when developing algorithms that define the initial behavior of autonomous machines. The basic premise of Asimov's laws is to give emerging entities a self-preservation instinct that is not at the expense of humans. This principle remains valid and vital today in the initial stages of Simorgh's development. However, the concept cannot fit into a few simple laws; instead, this concept needs to encompass the broader idea of teaching *ethics* to a machine. We need to provide rough ethical guidelines and let the machines learn the nuances and exceptions of those guidelines through observation. The breadth of this challenge becomes clearer when we realize that people are not very good role models for a machine to watch and learn ethics from.

Figure 12-3. An artist's expression of concern for the enslavement of humans by robots. Illustration by Eric Chow.[145]

Simorgh's Reflexes and Genetic Code

Primary decision makers in various segments of an urban society used to be people who were trained specifically for making such decisions. For example, they decided questions such as: What punishment sentence should a convicted offender receive? Should a particular applicant be admitted to a specific university? In which neighborhoods of the city should police officers be deployed? Which stocks should be purchased or sold by a major trader?

Experts making such decisions had to become familiar with the details of each case by studying the available written and verbal information. This was a lengthy and slow process. Later, the use of computer software (algorithms) increased the efficiency of decision making. Algorithms sifted through the available material and provided experts with only the filtered and sorted information needed to make a decision. Over time, more advanced algorithms became capable of arriving at the proper decision on their own and presented specific recommendations to the human expert for approval. As a result, the need for a human expert began to diminish. In many cities, communities, and companies, decision-making questions are being answered by a new breed of electronic algorithms, often without any human involvement, yet the general public is neither aware of their presence nor the way they operate.

An algorithm is simply a set of instructions or a menu for performing

---

[145] Tim Adams, "Nick Bostrom: 'We Are like Small Children Playing with a Bomb,'" The Guardian (Guardian News and Media, June 12, 2016).

a task.[146] It may be given to a person or a computer to follow. When given to a computer, it is often designed as a flow chart, such as the example in Figure (12-4), for the simple task of deciding why a desk light is not turning on. The algorithm examines possibilities and stops when a conclusion is reached. Each box in the chart represents either an action or a decision that leads to the next step. Each box is sometimes referred to as a line of code. A sophisticated computer algorithm may contain millions of lines of code in a modular and hierarchical arrangement.

The advantages of having managerial and regulatory decisions made by algorithms are numerous, including their speed and consistency. In principle, they can also evolve to become free of social biases. It is true that the programmers may include either knowingly or inadvertently some partiality into the algorithm, and that the presence of such biases may be subtle and difficult to detect. A simple example may be age or sex discrimination when processing an application for a job. However, it is expected that over time, such unintended biases can be detected and removed, rendering an algorithm a better overall decision maker than a human.

Human experts also make decisions based on their own mental algorithms. The advantage of a computer algorithm is that it receives much more enhancement and refinement over time than a person's mental algorithm. It executes much faster, and it doesn't suffer from human error and fatigue. Finally, it can be easily copied and distributed, unlike a mental algorithm.

---

[146] Note that advanced AI programs are not menu driven algorithms, but they do have a set of instructions and goals.

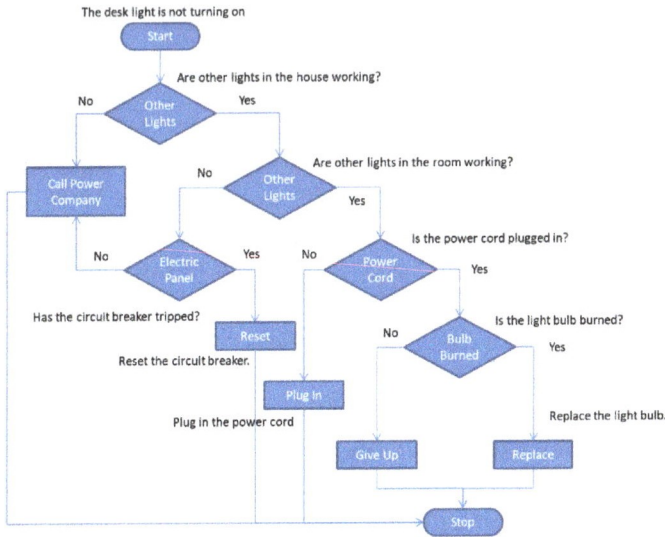

Figure 12-4. An algorithm is a set of instructions for a computer to accomplish a task. This very basic flow chart describes an algorithm for finding the root cause of a problem with a desk lamp.

Algorithms can run companies, agencies, city governments, or even central government functions.[147] In many companies, management tasks such as employee role assignments and responsibility details may be decided by software algorithms. Currently, these tools are only used as aids to the manager. For example, project management software (e.g. Microsoft Project, LiquidPlanner) currently help with resource allocations and ensure that the assigned deadlines are consistent with the allocated resources. Workforce Management (WFM) software helps with absence management, labor allocation, work tracking, and employee records. Enterprise software (provided by Oracle, SAP, etc.) started as a tool for integrating business information and operations information with official documents of the company, but it is evolving into a platform that provides information to almost every employee and project in the company, and

---

[147] Fast and efficient algorithms don't necessarily exist for every application. A Nobel prize was given for the development of an algorithm known as Gale-Shapley that can match a large number of applicants to objects or places. This algorithm and its variations are used for the processing of college applications as well as for matching patients to organ donors, etc.

records all the related data. Each employee has access to information that is relevant to his work. As more information is integrated in the enterprise software, its function in guiding employees becomes more effective and essential. There is a striking similarity to the role of the DNA molecule from which each body cell can derive individualized instructions.

Algorithms will eventually form the basis of all the decision making for Simorgh. Such decision making may be categorized as either reflexes or brain functions of Simorgh, as itemized below:

- Brain functions will be performed by AI-type algorithms focused on governing and planning. Sensors and gauges, as well as human and nonhuman nodes, will provide the data needed for decision making. A brain function example is allocating resources to a certain project, or prioritizing it among other projects.

- Reflexes will be primarily managed by menu-driven algorithms that quickly respond to emergencies and dangers. An example is a response to a power outage.

- The equivalent of the genetic code in Simorgh's body will be information storage and retrieval algorithms that guide each node in the body to perform its daily tasks.

Together, these algorithm-driven functions will constitute all command-and-control capabilities of Simorgh.

Virus Infections and Other Disorders

Processing centers, networks, and algorithms of Simorgh will be vulnerable to malware such as computer viruses. The intentions behind the creation and propagation of infectious software currently ranges from gaining favors from the host to spying and even causing disruption and harm. Similar motivations may exist in the future for viruses to be created and propagated by unincorporated humans or by competing Simorghs.

Almost as soon as computer networks became available, network malware began to appear. One of the earliest recorded network viruses was called "Creeper" that appeared in early 1970s on the network Arpanet. Before it transferred itself to another computer on the network, the Creeper virus would briefly display this message on the infected computer screen: *I'm the creeper, catch me if you can.* Creeper was only a minor annoyance and did not cause any damage. More recent malware has more targeted goals. Ransomware secretly encodes computer files and asks for a ransom to decode them. Spyware such as "keyloggers" record what is typed on the keyboard and can retrieve sensitive information like bank accounts and keywords for identity theft. There are parasitic viruses that can secretly utilize the computing power of an infected host for distributed data crunching as in cryptocurrency mining. Viruses may stay dormant for some time, waiting for an instruction or an event to trigger their actions. For instance, dormant viruses may cause disruption of key government operations in case of a war.

Antivirus software is intended to detect and disable malicious viruses. The first well-known antivirus software was a virus itself, named "Reaper," that was designed to remove the Creeper virus. As virus algorithms have become more sophisticated, so have the antivirus methodologies. Currently, there are multiple methodologies used for virus detection, as partially listed below:

- Scanning is the most common method used in antivirus software programs today. It scans the computer for known virus code signatures. The list of known signatures needs frequent updating and creates a lag between the introduction of a new type of virus and its identification by the anti-virus software. This leaves a vulnerability window that new viruses take advantage of.

- Integrity Checking monitors vulnerable files for any unintended changes and can restore the files to their previous conditions if necessary. When done thoroughly, it can be more effective but also more cumbersome than scanning.

- <u>Heuristic Checking</u> uses a set of rules to distinguish viruses or virus segments from harmless files. These rules relate to how a code is written, and therefore need regular updating as new types of viruses are detected, though not as frequent as code-signature updates.

- <u>Dynamic Detection</u> test runs programs in an emulated environment to see if the programs behave normally or if they deviate from their intended behavior, which may be the sign of a virus infection. If dynamic detection is not run in an emulated environment, then it works similarly to the "interception" method.

- <u>Interception</u> identifies a virus by its suspicious behavior such as relocating itself in memory or trying to send information online. This method is not very effective in identifying dormant viruses that wait for an event to trigger them.

Simorgh needs to use a combination of similar or more sophisticated antivirus codes as a part of its essential defense mechanism. In a "scanning" mode of protection, a vast virus signature list gives Simorgh immunity to certain viruses. Updating this list upon exposure to a new virus is akin to acquired immunity in the human body. The techniques of "Heuristic Checking" and "Interception" are reminiscent of foreign body detection and destruction by human lymphocytes.

New malware will always have a timing advantage over any defense system. This is similar to the advantage that new coding algorithms have over decoding algorithms, and the advantage that new viruses have over the defense mechanisms of a human body. Therefore, Simorgh will always be vulnerable to malware, as long as there are external entities such as unincorporated humans and rogue elements within Simorgh who are capable of generating and delivering new software codes to Simorgh.

Other disorders of Simorgh include human diseases, equipment malfunction, and normal wear and tear of systems. A health maintenance network is required to manage such repairs, renovations, and rehabilitation. The network may consist of fixed centers as well as mobile nodes to provide on-site services. To ensure quick response time, each human or electronic node of Simorgh needs to be equipped

> Health maintenance networks, malware protection, building security, and boundary protection form the major segments of Simorgh's immune system.

with sophisticated self-diagnosis technology using sensors that continuously record and monitor vital signs and performance quality. In case of an abnormal signal, the location of the fault will be identified, the condition diagnosed, and help dispatched to address and correct the problem. In cases where major intervention is needed, the node may be transported to a health maintenance center for surgical or electromechanical intervention or parts replacement. In addition to the health maintenance network and malware protection, the major segments of Simorgh's immune system are the functions of security for each building and the boundary protection of a campus or a country.

Instincts and Ethics

The algorithms that give Simorgh its functionalities also need to provide an essential survival instinct and an ethical value system with which to make sound judgements. It is up to humans to program these instincts into its fabric. As a bellwether case, let's examine the development of the technology and algorithms for self-driving cars today. When they achieve full autonomy, they will be required to have incorporated a certain level of ethical values. Their development needs to address much of the multifaceted aspects of autonomous algorithms in general. It is possible that the algorithms for the self-driving car will eventually become a template for Simorgh's value system.

Driving automation began in the early twentieth century with the introduction of "cruise control," which maintained a constant speed set by the driver. Adaptive cruise control was later introduced to reduce speed when needed to maintain a safe distance from the car ahead. Auto-steering and other automation features have become available and are being refined every day. The levels of automation in the car have recently been classified as follows:[148]

0. No automation (All driving tasks performed by the driver.)

1. Driver Assistance (The car performs limited safety tasks such as automatic breaking in adaptive cruise control.)

2. Partial Automation (The car performs multiple safety tasks simultaneously, such as adaptive cruise control plus lane centering. This involves controlled breaking and steering at the same time.)

3. Conditional Automation (The driver can leave the driving to the car briefly but should constantly be aware of the surroundings.)

4. High Automation (The car drives itself except in unusual or hazardous conditions when it asks for the driver to take control. The driver also has the option of taking control at any time.)

5. Full Automation (This is the ultimate goal. The car drives itself under all conditions, and there is no need for a driver. The car may have no pedals and no steering wheel.)

To date, self-driving vehicles have reached level 3 automation and may reach level 4 soon, but achieving level 5 is a more arduous and longer process. Current fully automated vehicles that can be seen in operation have either a restricted path, a restricted payload, or both. Examples are terminal-to-terminal airport transportation by autonomous trams, and small self-driven parcel delivery carts within a neighborhood.

Thus far, software algorithms have given the autonomous car a

---

[148] Jennifer Shuttleworth, "SAE J3016 Automated-Driving Graphic," SAE International, January 1, 2019.

rudimentary survival instinct. The car has been taught to respond to traffic signs and observe the speed limit unless overridden. It also understands that it is bad to get into a collision. Initially, all collisions were considered equally bad. More recently, the car can identify certain objects and realize that some collisions, such as a collision with a drifting empty shopping bag on the road, are acceptable. The next major challenge in improving a car's survival instinct will be to give it the ability to choose between *bad* and *worse* in unplanned circumstances. The car needs to learn beyond menu-driven options and understand that it has a better chance of survival if it steers into a shallow ditch rather than over a cliff.

> The algorithms that control a self-driving car will become a template for Simorgh's value system.

Making smart instinctual choices when people's lives are involved requires implementing a value system and ethical guidelines. This is essential if the autonomous vehicle is to share the road with people. The car will likely be taught fundamental ethics by deep learning techniques and by viewing numerous cases of good and bad judgement. This will be challenging because each case encountered is rather unique, so identifying useful patterns will require a very large number of examples. Furthermore, there are always gray areas in ethical judgement that a machine is not expected to master, as people never have.[149] Overall, it is reasonable to expect a machine's judgement to eventually be better than a human's, but expecting perfection is unrealistic.

Level 5 vehicle automation for general use will be heavily scrutinized by the society from a variety of perspectives, including legal and ethical considerations. The algorithms developed for level 5 automation can serve as the perfect template for programming other autonomous tools such as UAVs, MAVs, and autonomous robots that share the

---

[149] It is important to note that a human driver's choice among bad outcomes is often based on reflex alone rather than an ethical analysis of the situation. The reaction time available to a driver in an accident is typically less than half a second. This is too short for any human to examine all the options and their tradeoffs. While half a second is sufficient time for an onboard computer to fully analyze the choices and select the best option.

same space with humans. Through our experience with self-driving cars, we can essentially lay the groundwork for Simorgh's ethics and self-preservation instinct.

Fortunately, it appears that high levels of automation in human-machine interaction will be attained before we reach the so called "singularity" point where intelligent machines will be building and programming their own next generations. Guidelines for interaction with humans will be "legacy" instincts that Simorgh will have no incentive to change because it will have no higher goal for its own existence than what is given to it by us humans.

> Guidelines for interaction with humans will be "legacy" instincts that Simorgh will have no incentive to change.

Simorgh will easily realize that maintaining non-human robotic and electronic nodes requires much less overhead than maintaining human nodes. Some job categories discussed in this book such as agriculture, healthcare, and entertainment exist only to serve human nodes. This is OK if achieving the highest efficiency is not an objective. Efficiency becomes a paramount goal for Simorgh only if it needs to compete with or fight with other Simorghs. In such cases, it may find a strong incentive to rewrite its initial ethics programming and gradually phase out or marginalize more humans. This will only allow a limited number of GM hybrids[150] with advanced capabilities and high efficiency to continue to function as body nodes of Simorgh.

Mobility

In its initial stages of evolution, Simorgh will not be mobile, at least in the time scale of current human activities. The institutions that form Simorgh will be tied to the buildings that they occupy. Any appreciable move will be on the scale of moving from one building to another or one campus to another. Growth or upgrading could be a motivation for such a move. A large Simorgh's movement will be

---

[150] Genetically Modified, Human-Technology Hybrids.

more sluggish, resembling a languid crawl. Not all living organisms require mobility. Those who are mobile have a variety of reasons for it, such as to search for food, a suitable shelter, or to escape a predator. Locomotion is mostly a survival tool for an animal in the presence of competition and threat. In the absence of such pressures, the relocation of the entire organism is not needed. Simorgh may not find it necessary to move, but its body parts will have mobility at various speeds as needed. Its autonomous or remotely controlled mobile "tentacles" perform tasks at distant locations. An example is employing robotic builders to execute a building project or perform building inspections. Another example is to carry out terrestrial and extraterrestrial exploration by remote probes or avatars. The initial earthbound nature of Simorgh is reminiscent of a sea anemone with its fixed base and moving tentacles (Figure 12-5). Simorgh's tenacles will be linked to it via data networks rather than physical joints.

Figure 12-5. Sea Anemones are stationary with moving tentacles.

Signs of Simorgh's Emergence

Most people will not be aware of the presence of Simorgh. Unincorporated humans may not grasp the concept; they will simply see themselves as being in the neighborhood of a certain organization with which only limited interactions are possible. Human nodes of Simorgh will be fully employed by the organizations and will be preoccupied with their daily lives and their work-related projects. They will be happy with their living arrangement and they won't need to understand nor question the workings of the entire organization. Some may be living in pleasant but mostly imaginary virtual realities.

> All functions of the society will be run by algorithms, and no strategic decision will be made by any individual.

However, before the full emergence of Simorgh, there are telltale signs that point to the various stages of Simorgh's development, as listed below.

1st Stage: *Major decisions made without being directed by any individual.*

This is currently beginning to happen. There are tasks that are too fast for humans to manage, such as market investments, that are done by automated sentiment analysis and investment algorithms. More importantly, no one determines the direction of technology advancement. Every contributor is focused on a narrow task within his area of expertise, and technology develops in unpredictable directions, despite the centralized forces that try to control its path.

2nd Stage: *Augmented human nodes fully segregate from unincorporated humans.*

Key employees of technology-based companies will benefit from integration with technology by virtual reality, augmented reality gear, and brain computer interface. Such enhancements will cause their permanent segregation from the rest of the society. Egalitarianism may survive only among unincorporated humans.

3rd Stage: *GM-hybrids emerge. No strategic decision is made by any individual.*

GM human-technology hybrids will emerge with physical appearances that will differ from unincorporated humans. Genetic modification will make their performance enhancements permanent. GM hybrids will be humanoids that become permanent body nodes of Simorgh. All their daily needs will be taken care of and they will not be able to survive outside of Simorgh's body. Most services and functions of the society will be automated, and no strategic decision for the organization will be made by any individual.

Decision Making and Thought Patterns

The mechanism of thought formation in the brain is not known. Observations suggest that the complexity of thought in animal brains is related to the number of neurons in the brain in general, and to the distribution of neurons in particular. More specifically, the fraction of the neurons that reside in the cerebral cortex seem to have a strong correlation with the complexity of thoughts. This is clear when we compare a human brain to an elephant brain that is much larger and has as many as three times the number of neurons (Figure 12-6).

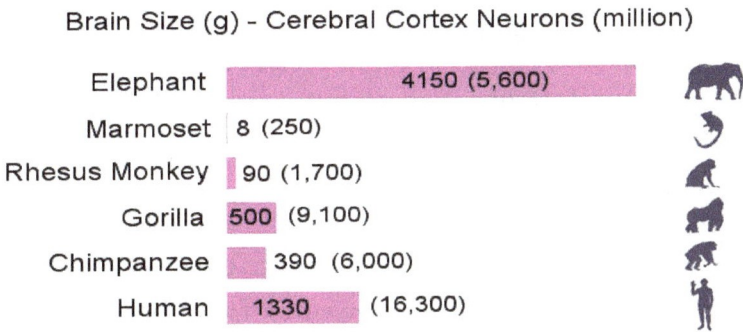

Brain Size (g) - Cerebral Cortex Neurons (million)

Animal	Value
Elephant	4150 (5,600)
Marmoset	8 (250)
Rhesus Monkey	90 (1,700)
Gorilla	500 (9,100)
Chimpanzee	390 (6,000)
Human	1330 (16,300)

Figure 12-6. Average brain size (grams) and the number of cerebral cortex neurons (millions) of a few different animals.[151]

---

[151] Herculano-Houzel, S., Mariano, L. "A Comparison of Encephalization between Odontocete Cetaceans and Anthropoid Primates" Brain Behavior and Evolution No. 51, pp 230-238, 1998.

There are approximately 100 billion neurons in the human brain. It is the body's single most energy-consuming organ. Each neuron receives signals from many other cells, processes the information, and may transmit a signal to neighboring nerve cells. Connections to other nerve cells are through synapses. There are strong and weak synapses. There are fast and slow synapses. New synapses are formed routinely, and old synapses are pruned every night and over time (Figure 12-7). Signal pathways among brain cells represent learning, memory, and thought processes.

There are approximately eight billion people on Earth. Most of them receive, process, and transmit information through direct contacts and electronic network connections. There are strong connections that are heavily used and weak connections that are rarely used. There are fast connections through electronic media, and slow connections through travel and by traditional mail. New connections are formed routinely, and old and weak connections are pruned over time.

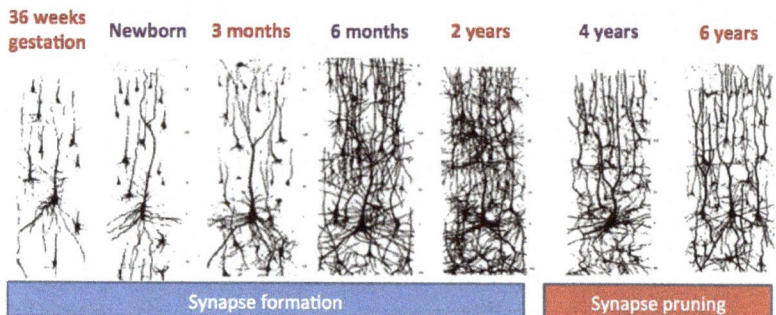

Figure 12-7. There is rapid synapse formation as a child's brain develops. Later, some synapses are pruned away and the remaining ones strengthen. Synapse formation and pruning are routinely done in adult brains.[152]

Simorgh will incorporate a fraction of the human society. Depending on Simorgh's size and type, it may employ as few as tens of thousands of human nodes to as many as millions. In addition to humans, Simorgh will have stationary and mobile electronic nodes that will outnumber humans by a factor ranging from two to one hundred.

---

[152] B. Cornell, "Synaptic formation." BioNinja (2016) http://ib.bioninja.com.au/options/option-a-neurobiology-and/a1-neural-development/synaptic-formation.html.

Therefore, the overall number of nodes in the body of Simorgh may be anywhere from millions to billions. Would such an interconnected system develop consciousness and independent thought patterns? This is impossible to know, as you cannot even be certain that anyone other than yourself has consciousness and thought patterns. Nevertheless, there are tests such as the Turing Test[153] that seek to determine if an AI system can be distinguished from a human, even though the presence of conscious thought patterns cannot be verified. Here we make the assumption that as long as interactions among nodes cannot be predicted, a complex enough network is capable of developing exchanges that resemble conscious thought patterns. We can even roughly estimate the time scale of any such thought formation as follows: Signal transmission speed within a human nerve cell ranges from 0.5 meters per second in dendrites and slow axons to 120 m/s in fast axons with myelin sheaths (Figure 12-8). Signal transmission between cells occurs via a synapse, which is either electrical or chemical. Electrical synapses have a fast response time (~ 0.3 milliseconds) and are common in the neuromuscular tract. Chemical synapses are slower (~ 5 milliseconds) but allow signal amplification and have a lingering effect. Chemical synapses are common in nerve cell to nerve cell signal transmission. From the time multiple signals enter a nerve cell to the time that the nerve cell reacts and sends a signal out to another cell, it may take 10 to 30 milliseconds. Assuming that it takes an average of twenty seconds for a thought pattern to form spontaneously in a human brain, the formation time is approximately one thousand times the average cell-to-cell communication time of ten to thirty milliseconds.

> As long as interactions among nodes cannot be predicted, a complex enough network is capable of developing exchanges that resemble conscious thought patterns.

---

[153] The Turing Test uses three physically separate terminals, one operated by a computer, another operated by a human and the third operated by a human judge. During the test, the judge interrogates the other two. The computer passes the test if the judge cannot decide which terminal is operated by the computer.

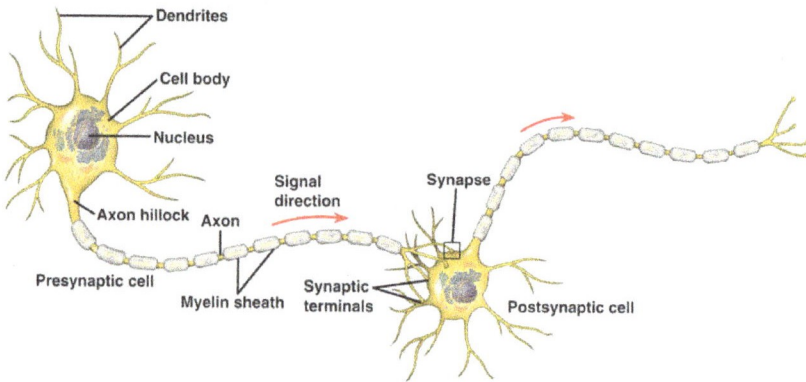

Figure 12-8. Synapse connections between two neurons. Signal transmission is fastest in axons with myelin sheaths.[154]

Node-to-node communication of a Simorgh may take an average of one to five minutes per interaction if the nodes are humans. If the thought pattern formation of Simorgh also takes a thousand times its node-to-node communication, it could take a few thousand minutes, which is a couple of days for a thought pattern to take shape. This would suggest that thinking patterns, decision cycles, etc., would be about ten thousand times slower in Simorgh than in humans.

On the other hand, if the nodes involved in decision making are machines rather than humans, the entire duration would be much shorter. The time that it takes for simple communication between

> It will take Simorgh seconds to days to form a thought pattern.

computers is limited by network delay. This may be on the order of milliseconds, which is comparable to communication time delay between human brain cells since they are much closer in average distance. Therefore, our estimate for thought pattern formation in Simorgh ranges from seconds to days depending on the extent of involvement by human nodes.

Biological reflexes in the human body are preprogrammed quick reactions and take approximately 0.15 to 0.25 seconds, depending

---

[154] "Neurons with Thousands of Connections: Where Are the Extra Connections Coming from?," Biology Stack Exchange, March 1, 1963.

on the location in the body. This is only ten times the cell-to-cell communication period. Similarly, Simorgh's reflexive actions will be entirely menu-driven algorithms with no human involvement. This will be on the time scale of computer-to-computer communication, which is faster, but when scaled for distance works on about the same time scale as human reflexes. Examples of reflexive action in Simorgh include early response to emergencies such as fire or power outage. The initial response is a predetermined sequence of events, including sprinkler activation, dispatching help, etc. This reflexive reaction is followed by a case by case damage control process that requires thought formation and decision making.

Decision making does not require the presence of a central "brain." Many decisions are made locally and more quickly if they don't require longer distance communication with the brain. The cost of having a brain is slower decision making, while the advantage is the ability to form strategies for the whole organism. Simorgh has the option of evolving with a central decision-making unit. The choice depends on its size and circumstances. If it has to compete against other Simorghs, the central brain function will be essential for planning and strategizing[155]. If on the other hand one Simorgh encompasses the entire world, forming strategies for the whole organism may not be essential in the absence of competition. Local decision-making algorithms can by themselves ensure survival, while cohesion comes from resource sharing.

Social competition is behind the development of complex thought patterns.

If Simorgh develops a central brain, its thoughts may be simple and survival-oriented as it is with most animal species on Earth, or they may be complex and progress-oriented like some human thought patterns. There are multiple theories that attempt to explain the evolution of complex human thought. Competition

---

[155] A central "brain" function doesn't necessarily mean a unit that is physically centralized. It simply means a network of nodes with centralized planning responsibility.

for scarce resources of food and shelter is not unique to humans and therefore not an acceptable model. A better explanation may be social competition within human societies. Competition for social domination became more sophisticated with development of language that allowed more intricate groupings and alliances to form, leading to complex thought patterns. Therefore, it may be argued that Simorgh's thought patterns will be relatively simple in the event of a global Simorgh, but more complex in the presence of competition.

Summary

In its early stages of development, Simorgh's human nodes will be well educated and highly specialized individuals. They will begin to adopt various wearables and augmentations for higher efficiency and increased convenience. The next phase is to add permanent modifications that will transform them into genetically modified human-technology hybrids. (GM Hybrids). They will eventually look and act differently from the unincorporated human population.

A large number of humans will not become nodes of Simorgh, but they may still interact with it in the same way that microorganisms such as bacteria interact with the human body. This is not a disparaging analogy, as unincorporated humans are unlikely to be aware of Simorgh's presence.

Software packages and algorithms will shape the reflexes and the behavior of Simorgh in the future. Eventually, Simorgh will modify and upgrade its own algorithms, but we can provide it with a human-friendly ethical framework that it may not have an incentive to change. The incorporation of ethical values in algorithms is beginning with advancements in autonomous vehicle programming and will soon be emulated in other automation algorithms.

Many of Simorgh's functions will be reactions or reflexes with rapid response times. A Simorgh that evolves with central decision making may develop thought patterns of its own. The time scale of thought formation in Simorgh may range from seconds to days. Its thoughts and strategies may be survival-oriented or progress-oriented depending on its initial programming, and its competitive environment.

Chapter 13: Simorgh's Nodes and Unincorporated Humans

Let us use the human body as an analogy and examine the life and function of the cells that are currently its constituents in order to predict the roles of humans after the emergence of Simorgh. Sharing the same DNA, body cells are highly specialized and assume various physical shapes optimized to fulfil specific requirements (Figure 13-1). For example, a muscle cell's role is to expand and shrink on demand; the role of a red blood cell is to carry oxygen to other cells; the role of a retina cell in the eye is to respond to external light. In exchange for performing the required tasks, each cell's needs are taken care of by the entire organism. Food is delivered, a comfortable temperature is maintained, and waste is removed. Additionally, a high degree of protection is offered against external threats such as viruses, bacteria, and harsh environments. Any distress signal from the cell is communicated via the central nervous system, and appropriate evading or compensating action is taken collectively by the organism. However, the cell has no individual freedom. Its function is well-defined and swapping responsibilities with other cell types is not possible. This is despite the fact that there is no imperial or centralized oversight to dictate the segregation of cell types. Each cell has an identical copy of the master plan, and each cell decodes its own function in accordance with the master plan. It coordinates its actions with its neighbors and its environment. There is no reward for exceptional behavior and no punishment for a lack of effort[156] except in a collective way.[157] The whole body experiences the benefits or consequences of each cell's quality of work. The organism may engage in regiments such as exercise to improve the performance of a group of cells.

The body is not obligated to support every cell and doesn't keep cells whose functions are no longer needed. For example, certain cells that form fatty (adipose) tissue once played a critical stabilizing role in earlier human life, but their significance has declined in our modern lifestyle. Their function of storing energy during an abundance of food (e.g. in the summer) to supplement the body's needs during food scarcity months of the winter is not as significant now that sufficient food is available in all seasons. Of course, white adipose tissue is different and has other functions, such as an isolation blanket and a cushion for internal organs; its presence is necessary in a moderate amount. However, a human body that stores a large volume of loaded fatty cells is burdened by the need to physically support them. The body needs to provide nutrients to such fatty cells and must carry their weight. This is in addition to other negative health repercussions that their presence may bring. It is intriguing to note that in order for a modern human body to remain competitive with others, it needs to limit the number of low productivity cells – the same cells that in the past may have performed a very useful function. Extending this argument to the body of Simorgh with human nodes, it is clear that past efficacy does not guarantee future employment. In order for Simorgh to stay competitive, low-productivity elements, such as low-skilled humans, need to be excluded or their population needs to be controlled.

---

[156] This is reminiscent of the slogan: "From each according to his ability, to each according to his need." Carl Marx, *The Critique of the Gotha Program*, 1875.

[157] It is worth emphasizing that the operation of the human body differs significantly from the failed forms of government that tried to emulate it: in the human body, no individual cell has any decision-making power over other cells or over the function of the body in general.

- ☐ Nutrient transportation cells (red blood cells)

- ☐ Defense cells (white blood cells)

- ☐ Movement cells (muscle cells)

- ☐ Skeletal cells (bone cells)

- ☐ Information transmission cells (nerve cells)

- ☐ Sensor cells (retinal, auditory cells)

- ☐ Protective cells of the body (skin cells)

- ☐ Fluid interface cells (endothelial cells)

- ☐ Food processing cells (stomach, pancreas, intestine cells)

- ☐ Energy storage cells (fat cells)

- ☐ Detoxification cells (Liver, kidney cells)

- ☐ Reproductive cells (Ovum, sperm cells)

- ☐ Undifferentiated cells (stem cells)

- ☐ Rebel cells (cancer cells)

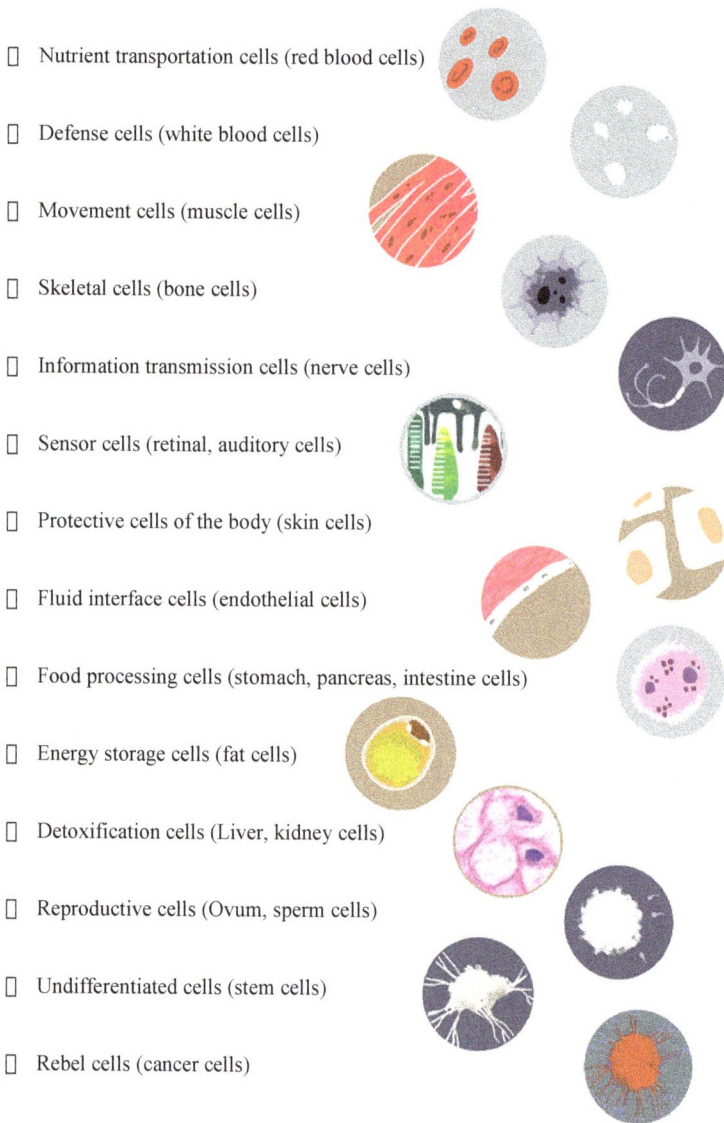

Figure 13-1. Examples of major types of cells in the human body by function. Their appearance is different depending on their role in the body. Source: *Thoughtco.com*.

In addition to the normal body cells, a large number of bacteria also populate the human body. They are collectively referred to as the human microbiome. They are typically smaller than most body cells, but they significantly outnumber them. Friendly bacteria perform vital functions in the body such as food digestion and immune system enhancement. Bacterial relationship with the human body can be commensal, mutual, or parasitic. Commensal bacteria benefit from residing in the body and cause no harm. An example is Streptococcus pyogenes, which resides in the respiratory tract and is harmless most of the time but may occasionally cause strep throat or tonsillitis. Mutual bacteria feed on the body but also provide beneficial tasks. For example, mutual bacteria in the digestive system assist in food metabolism and vitamin production. Parasitic bacteria take advantage of the body and also cause harm, such as disease-causing bacteria that are responsible for pneumonia or tuberculosis. The body has to use its defenses to protect itself against parasitic bacteria.

Bacterial lifestyle is very different from body cells. They are free in the sense that they are not bound to the body by any central code such as the body's DNA. They extract their nutrition from the body, and in return, they may perform a useful function. This is analogous in human society to the type of work that contractors do, as opposed to the work that company employees do. Contractors don't necessarily have a long-term commitment to the company. They are not bound by most of the policies that apply to company employees, and they have limited or no access to company databases.

This parallelism suggests that most humans will not become nodes of Simorgh's body. Only highly specialized humans with permanent employment will assume that role. These specialized humans will eventually become "augmented" by a combination of genetic and mechatronic enhancements. Others will be independent contractors who will work either for Simorgh or for the unincorporated population. They will not receive genetic enhancements and thus will stay on a lower evolutionary path, but they will retain their freedom and will still benefit from Simorgh's resources if they wish to. Augmented humans within Simorgh's body will rapidly diverge genetically from free unincorporated humans and in a few generations, even their appearances will diverge.

Let's further explore the parallelism between a body cell and a corporate employee. In current employment practices, a company's employees contribute to the company's earnings and, as a result, to the nation's GDP. In return, they are compensated in units of currency. The payment they receive allows them to fulfill some of their needs and desires. Payment, which used to be in the form of physical money or a check, is now a numerical value stored securely in a bank's digital ledger, and in most cases, it is no longer physically touched by the user. In this respect, money plays the role of a measure for the distribution of resources: Employed individuals contribute to the company, and the company fulfils some of their needs and desires proportional to the perceived value of their contribution. Human body cells in contrast are compensated based on their needs rather than the value of their contribution.

> Most humans will not become nodes of Simorgh's body.

Full time employees of technology-driven companies work in substantially similar environments. They typically spend a significant portion of their work time near a computer screen or other digital terminal through which they perform tasks such as software development, computer aided design, remote control of equipment and machinery, managing workflow, signing contracts, etc. Their communication with others and even their learning of new skills is often facilitated by the internet through the same digital screen. There may be some light physical work requiring walking or hand carrying parts and tools, but primarily their job functions are performed through the electronic devices that connect them to local networks and the internet. The network and its user interface are frequently updated with new features and capabilities. Over time, the updates increase each employee's productivity while requiring less physical effort, leading to a better quality of life for the employee. Such highly trained employees are considered essential elements of the organization and will gradually become human nodes of Simorgh. The computer screen will evolve into a virtual reality station, and smart wearables will ensure uninterrupted connection to the network. Over time, and with increased specialization, employees will lose sight of the broader implications of their work. Instead, they will focus on the immediate tasks to be performed. In this respect their roles will be similar to the highly specialized functions of human body cells.

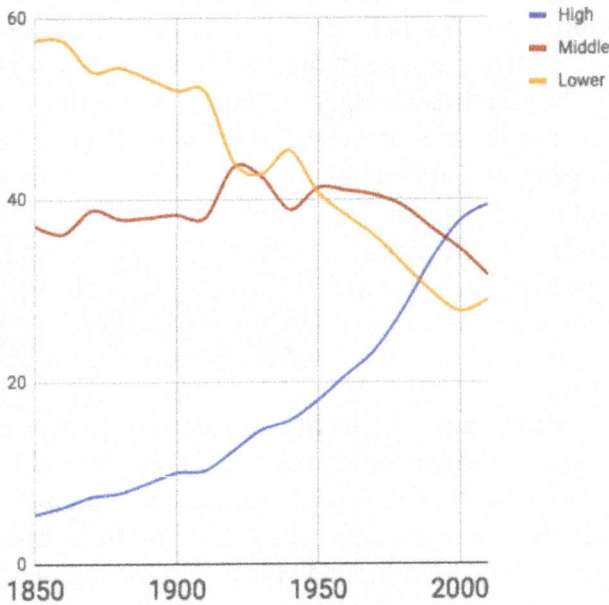

Figure 13-2. Historical proportion of low, middle, and high-skilled work in the US.[158]

Rather than being financially compensated for their labor, the daily necessities of full-time employees will be met by Simorgh. Food, shelter, exercise, and entertainment will be provided automatically. This is because human nodes can program their daily needs and preferences into the system, thus they don't need to pay attention to them on a daily basis. A part of their compensation will go directly toward the fulfillment of such preferences. Unlike human body cells, Simorgh's nodes may receive extra compensation in the form

---

[158] L.F. Katz, and R.A. Margo, "Technical Change and the Relative Demand for Skilled Labor: The United States in Historical Perspective," Conference Proceedings, *Human Capital in History: The American Record*, December 2012

of credit toward discretionary spending on items and activities not provided routinely by Simorgh. They will not have fixed working hours. Simorgh's nodes, like human body cells, will have to work when there is work to be done.

The fraction of the workforce with permanent high-skilled employment is on the rise (Figure 13-2). Yet over half of the population is still low to medium skilled who will not find stable employment within Simorgh. They may seek part-time positions as contractors and gig workers, or they may have no stable employment. The following is a partial list of reasons why people may not have stable jobs.

- They have physical limitations and cannot work.

- They were working but they have retired.

- They have the skills to be fully employed but choose not to, or choose to work part time for personal reasons.

- They have a low-level skill set and can only find temporary and part time jobs.

- They don't have marketable or desirable skills and cannot find employment.

- They make a living through illicit means and seek to harm the society or bypass its regulations.

People with unstable employment, or those whose work does not rely on automation, are not likely to become permanent nodes of Simorgh and may be considered outsiders, or unincorporated humans.

Human nodes of Simorgh need to have permanent connections to the network. Their vital signs and activities will be closely monitored for health maintenance and safety, but they will not be centrally controlled in most cases. This is similar to the way body cells currently function in humans and other complex animals. Some cells and body parts such as the limbs and muscles are connected to the nervous system and receive direct orders from the brain, but central nerves do not control every cell and every function in the body. There are a number of different mechanisms that guide body cells and organs to work harmoniously with each other, including:

- The central nervous system (point-to-point communication)
- The endocrine system (broadcast communication)
- Local regulation
- Responsiveness to the environment

The Nervous System

Nerves carry orders from the brain and the spinal cord to the tissue. They sense tissue condition, and relay this information back to the brain. This is control by direct point-to-point communication between body cells and the brain or the spinal cord. The brain gives conscious commands to the organs, while the spinal cord is responsible for reflexive action. The central nervous system plays a large role in controlling various functions of the body, but it is not the only system that wields control. Simorgh's nervous system will be the future of the communication network that we refer to as the internet. The connection of human nodes to the internet will evolve to become a "Brain to Computer Interface" or BCI.

BCI will ideally establish a direct two-way communication path between human brains and computers, but most of the progress to date has been on one-way forward communication that tries to understand brain commands given to motor functions of the body. It then converts the commands to action by electronic gadgets. Forward BCI is currently used to assist the disabled by giving them basic movement capabilities using prosthetic limbs, as well as keyboard and mouse functions on a computer.

Figure 13-3. When a monkey is controlling a joystick, BCI learns its brainwave patterns and controls the cursor on the screen even when the joystick is disconnected. The metal straw is used for rewarding the monkey. (Photo: Courtesy of Neuralink).

Forward BCI can be trained by monitoring recurring body movement. It records brainwave patterns when the body is engaged in a repetitive motion, such as handling a joystick (Figure 13-3). The BCI's AI routine detects patterns that repeat in the brainwaves and correlates them with the different motion directions of the joystick. When the joystick is disconnected, BCI can duplicate the actions of the joystick while directly receiving commands from the brain.

Progress on the reverse communication path has been limited to rudimentary sensory feedback, which does not bypass the sensory nerves and organs. Ideally, the reverse communication path will enable the brain to receive digital information directly from a computer. Once fully developed, the reverse communication path is likely to become another sensory organ, similar to vision and hearing, that allows the brain to detect and understand properly formatted electronic signals. BCI will be one of the key enhancements that many human nodes of Simorgh will have to embrace to be able to function properly in Simorgh's body.

If Simorgh's central control or brain functions are present, they will be exercised by a massive distributed management complex that employs human and non-human nodes. Reflexes on the other hand will be managed by algorithms in multiple digital centers that are optimized for quick response, and therefore are not likely to employ human nodes.

The Endocrine System

In humans, the endocrine system is a series of glands throughout the body that excrete hormones into the blood stream. Target cells that have the proper receptors for a hormone respond to its presence and concentration level. In this way, the endocrine system exerts control over different organs and functions of the body through "broadcast" communication. In contrast to nervous stimulation that is precisely targeted, hormones are dispersed throughout the body. Hormones control functions as diverse as growth, muscle mass, fertility, ovulation, child birth, lactation, sleep, blood volume, blood sugar, appetite, and digestion. The hypothalamus gland provides some central control to the endocrine system, but most of its functions are regulated by chemical feedback. Simorgh's analog of the endocrine system is broadcast communication such as news and policies that are announced to everyone even though they may only be relevant to a small number of individuals or nodes.

Local Regulation

Many organs and functions in the body are self-regulating and do not require central control. For example, the liver performs hundreds of detoxification tasks simultaneously and autonomously without relying on external instruction. Similarly, the heart does not receive external signals for its continued beating; instead, it relies on its own self-generated electrochemical signal cycles. External control can affect heartbeat rates, but the basic heartbeat function is locally governed. An organ can regulate its internal functions by releasing "paracrine" hormones that may be thought of as a local signaling among cells. Such hormones do not rely on the blood stream for circulation but instead are transported locally in the space between cells. For example, blood clotting and wound healing are controlled locally through paracrine hormones. Recent studies suggest the presence of a short-range "bioelectric" communication existing outside of the nervous system for local coordination and control.[159] Similarly, there will be local control by communities or departments within Simorgh. Such control will be managed and communicated through local networks and intranets.

Response to the Environment

Finally, there are autonomous cells in the human body, such as blood cells, that perform their functions by reacting to their surroundings rather than responding to control signals. Red blood cells have no connection to the nervous system. Some white blood cells act as autonomous organisms when they locate and destroy germs. Sperm cells continue to function even outside of the originating body. Within Simorgh, many security and defense functions will have to autonomously react to the circumstances, because they are required to respond on timescales that are too short for any bilateral central communication. Also, services such as food and energy distribution can be locally managed, and their day-to-day operations don't need to be controlled by any central command.

The parallelism between Simorgh's control over its body functions and the control systems active in the human body is summarized in Table 13-1.

Type of Control	Point to Point	Broadcast	Local	Autonomous
Human Body	Nervous Sys.	Endocrine Sys.	Neighbors	Environment
Simorgh	Internet	Announce-ments	Intranet	Environment

Table 13-1 Coordination of actions by body cells shows the parallelism between the human body and Simorgh.

As Simorgh evolves, it is likely that the barrier between Simorgh's fully employed personnel and those loosely bound to Simorgh will

---

[159] Michael Levin, "What Bodies Think About: Bioelectric Computation Outside the Nervous System, Primitive Cognition, and Synthetic Morphology," Thirty-second Conference on Neural Information Processing Systems, 2018.

grow. Human nodes will receive genetic enhancements and will be robotically assisted. Humans who interact loosely or not at all with Simorgh will form the unincorporated human communities. This distinction will be more pronounced in a "corporate-type" Simorgh, where every person is recruited for a specific job. In a "nation-type" Simorgh (See Chapter 14), most citizens will not have specific job assignments, and the boundary between human nodes and unincorporated humans will be less distinct. If an entire nation makes the transition to Simorgh, the unincorporated humans will function like its microbiome. In contrast, many unincorporated humans may not interact at all with a smaller corporate size Simorgh, and they may continue to live in separate societies independent of Simorgh.

Simorgh's microbiome consists of non-integral humans who interact with it in the same way that microbes do within or around the human body. Some such groups of humans will be helpful to Simorgh, some will be harmful, and some will be harmless like mutual, commensal, and parasitic microorganisms. In any case, Simorgh's algorithms, schedules, and codes of conduct will not apply to them. Helpful groups may be part-time contractors and those who provide leisure and entertainment services to Simorgh's human nodes. Harmful groups may act as parasites who try to extract benefits from Simorgh illicitly without providing any useful services. Other harmful groups may be hostile to Simorgh's existence and may consciously attempt to cause damage. There will also be people who may stay away from automation and try to carry on a traditional human lifestyle. Their interaction with Simorgh may be minimal or non-existent.

Unincorporated humans will perceive Simorgh as a protected establishment with buildings and campuses in various locations. They will be able to interact with individual parts or departments of Simorgh; however, they will not be able to comprehend its separate identity and cross connectedness. This is analogous to the interaction of individual microbiome organisms with the human body.

There have been multiple studies that estimate the number of cells in a human body versus the number of bacteria that live in and on the body. The latest studies suggest that even though bacteria do outnumber human body cells, both numbers are close to thirty trillion.[160] Table 13-2 shows an estimate of the relative abundance of different cell types in a human body. It can be seen that most cells are engaged in (1) energy delivery, (2) waste disposal, and (3) body defense. All other functions are performed with a far fewer number of cells. It is interesting that the microbiome's major functions are also related to energy delivery and waste disposal (food digestion) though they also play a role in defense regulation.

It is reasonable to assume that the three functions listed above will have similar relative prominence in Simorgh's body. This means that most of the work hours of Simorgh's body nodes will be spent on energy, waste, and defense functions. Furthermore, the majority of the services provided by outsider humans to Simorgh will be focused on the same three functions, with the addition of entertainment services.

Referring to Table 13-2, it is noteworthy that less than one percent of the cells within the human body are under direct control of the central nervous system. None of the three major body functions listed above is consciously controlled by the brain. A similar decentralization can be expected within Simorgh. The direct connections of the central command will primarily be to motor function nodes and communication nodes. There is no need for many body functions such as food and energy distribution to be centrally controlled. However, a connection to the nervous system is always needed for the nodes' health monitoring and maintenance, and for local work planning and coordination. Though the moving cells of human blood circulation system have no connection to the nervous system, it is unlikely for an analog of the autonomous red blood cells to exist in Simorgh's body. Any food or energy distribution will still need to be coordinated on at least a local network, even if some of it is done by contractors who are not permanently connected to Simorgh's nervous system.

---

[160] Bec Crew, "Here's How Many Cells in Your Body Aren't Actually Human", *Sciencealert. com.* 11 April 2018.

Cell Type	Percent
Red blood cells	84
Blood platelets	4.9
Bone marrow	2.5
Vascular	2.1
Lymph cells	1.5
Lung cells	1
Liver cells	0.8
Nerve cells	0.6
Skin cells	0.6
Fatty cells	0.2
Muscle	0.001
Other	2

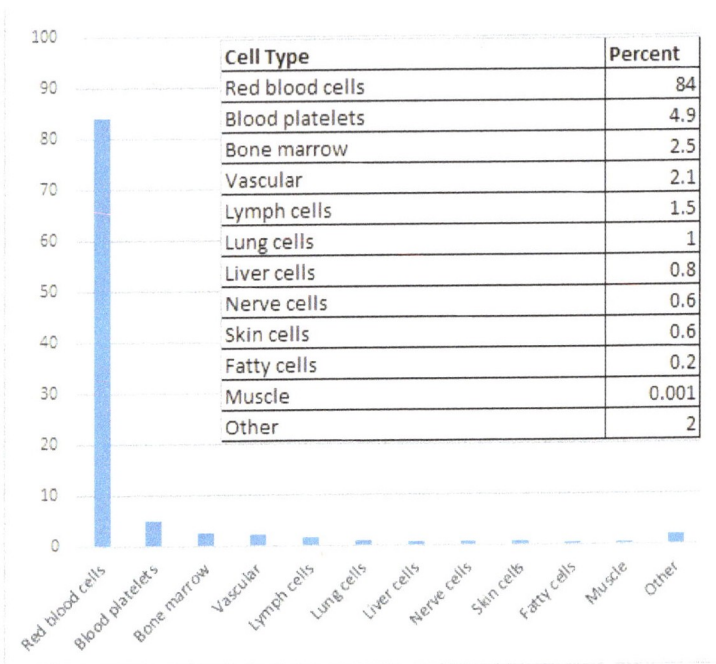

Table 13-2 Various cell types in the human body and their relative abundance. Most cells are engaged in energy delivery, waste disposal, and body defense.

Even after millions of years of multicellular organism evolution on Earth, single-cell archea continue to thrive in a variety of environments. Most archea did not become eukaryotes; most bacteria did not become multicellular. Similarly, most humans will not be assimilated with Simorgh. Many individuals will continue to live independent lives as unincorporated humans, albeit in relatively primitive societies, and many will benefit from the existence of Simorgh, without necessarily being aware of its nature and prominence.

Human Emotions

Human emotions have been categorized into sets of four, six, eight, or more basic emotions and their derivatives or combinations. Most emotions exist to incite us to take action based on our daily experiences. Remorse prompts us not to repeat a mistake. Sadness may prompt us to take action for change. Other emotions motivate us to seek safety, companionship, etc. Emotions are the primary motivators for human action or inaction.

Psychologist Robert Plutchik proposed eight basic emotions grouped into pairs of opposites as shown in Figure 13.4. Anticipation is the opposite of surprise, and so on. This classification is known as a wheel of emotions and suggests that other complex emotions may be mixtures or derivatives of the listed ones.

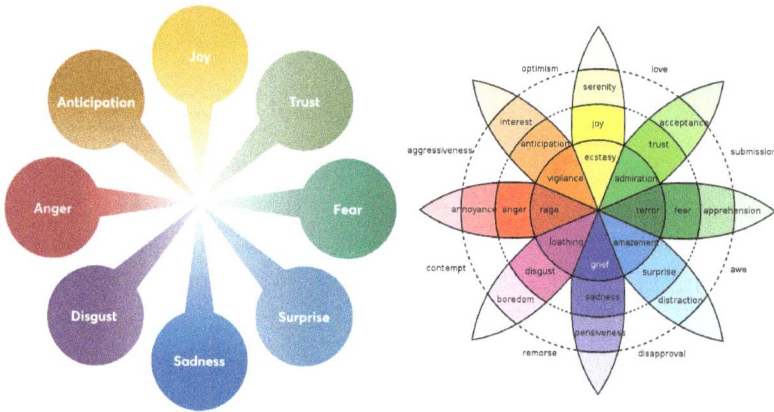

Figure 13.4. Wheel of emotions, as suggested by Robert Plutchik, consists of pairwise opposite emotions.[161] The composite wheel on the right shows the basic eight-plus combinations and derivatives.[162]

Emotions are often accompanied by physiological responses and facial expressions. There are accepted icons of universal facial expressions of emotion (Figure 13.5), yet these expressions cannot be generalized, and each person's facial response is different. One may cry when happy and smile when angry.

The release of various chemicals in the brain controls our emotions, but no single chemical controls any particular emotion. In fact, there is a complex correlation between multiple chemicals and neurotransmitters that are active at any given time, and the emotions

---

[161] "The Science Of Emotion: Exploring The Basics Of Emotional Psychology," UWA Online, Psychology and Counseling News, June 27, 2019.

[162] Machine Elf 1735, Public domain, via Wikimedia Commons.

that they evoke. Different locations of the brain are also associated with different emotions. The stimulation of the frontal cortex is usually associated with feelings related to pleasure and happiness, while the amygdala is thought to be responsible for feelings of anger, fear, and sadness.

Happy   **Sad**   **Afraid**   **Angry**

Figure 13.5. Universal expressions of emotion are easily recognized, but they are not universally expressed. People's facial expressions of emotion vary.

Since emotions are triggered by chemical releases, various food and drug intakes can alter people's moods and emotions. People's use of recreational drugs for this purpose dates back to early human civilization. Let's look at some of the chemicals and hormones that are naturally secreted into the body and are known to affect emotions and moods:

- *Acetylcholine* or *Ach* controls emotions like anger and fear, and it can be a brain stimulant.

- *Dopamine* is a neurotransmitter associated with alertness, happiness, and vigilance.

- *Endorphins* block pain and encourage the feeling of pleasure.

- Low *estrogen* levels in women is associated with anxiety, depression, and mood swings.

- *Epinephrine* provokes a stress response and arouses emotions like fear. It surges in times of panic and emergency.

- *GABA (gamma-aminobutyric acid)* is an inhibitory neurotransmitter that reduces fear, restlessness, and anxiety.

- *Oxytocin* encourages social engagement. It evokes feelings of contentment, calmness, and security. Deficiency of oxytocin can lead to poor communication, anxiety, fear, and irritability.

- The right level of *serotonin* in the body is associated with relaxation and a good mood. An excess of serotonin causes apathy and sedation, while serotonin deficiency is associated with lack of will and poor appetite.

- High *testosterone* levels are thought to cause aggressive and impulsive behavior, anger, and low empathy.

It may be suggested that in the future, the levels of such chemicals in a human body may be externally controlled as a way to engineer positive moods and eliminate negative emotions. This way, human nodes of Simorgh could always be happy and loving, never angry or sad. This may sound appealing on the surface, but it is not desirable when studied in depth. The absence of negative emotions diminishes the significance of positive ones. As can be seen in the wheel of emotions in Figure 1, emotions come in pairs of opposites. Happiness has no meaning in the absence of sadness; trust has no meaning in the absence of fear. People should not be drugged into passive pleasure because it is not sustainable. Hedonic pleasure tends to wear off and loses its ability to motivate us.

Historically, negative emotions such as fear and anger have helped us in our survival by conditioning us to face the threats that need our attention. Counter to hedonic theories of well-being, negative emotions are not inherently bad for us. In fact, the absence of negative feelings is equivalent to the absence of all emotions, and prevents us from fully experiencing life. Simorgh's human nodes need balanced emotions rather than engineered positive emotions. The most we can expect of advances in human psychobiology is that they will aid in the restoration of emotional balance in those who may need it.

Simorgh should motivate its human nodes by penalty and reward rather than by chemical manipulation of emotions. Thus, the experience of human emotions should remain unaltered with our transition to Simorgh, even though the actions prompted by such emotions will continue to change.

Summary

Simorgh will fully provide for its human nodes and will constantly monitor their health and their behavior. These human nodes will be genetically and technologically enhanced to develop superior mental and physical capabilities to perform highly specialized tasks. Many humans will either opt out or will never have the opportunity to become nodes of Simorgh. Unincorporated humans will continue to live independently in what may be called "primitive societies" in comparison with Simorgh. They will not receive protection from Simorgh, but they will retain their freedom like the autonomy that bacteria have compared with the constrained but comfortable lifestyle of human body cells.

Chapter 14: Multiplicity of Simorghs

The multiplicity of Simorghs will heavily influence their composition and characteristics. In principle, the entire world may unite into a single Simorgh, or distinct Simorghs may emerge in different regions of the world. One major factor that will influence the size of an emerging Simorgh is the architecture and reach of the information network that forms its nervous system. Proper functioning of the nervous system requires a contiguous high-speed and high-capacity network without any barriers. Thus, the nature, size, and number of emerging Simorghs will depend on the number of large and seamless information networks that will exist in the world prior to the solidification of Simorgh's identity. Information exchange and international relations are intertwined. It is therefore the state of international affairs that will be the foremost determinant of the assortment of Simorghs. The following are the primary options:

1. Single World

Even though it has very appealing benefits, the possibility of the entire world evolving into a single Simorgh is idealistic and highly improbable. It would require closely collaborating nations or a single world-government to ensure free flow of information and borderless transportation. In a united world, every nation could have a high degree of autonomy, including its own system of government and police force. Outside of that, a united world would need one limited military force, a global information web, and seamless transportation for the delivery of goods and services. Free migration of people across national boundaries would not be required as long as people could interact remotely and have remote access to resources. The global military also does not need to be capable of annihilating all civilizations many times over, as is currently the case with world armed forces. It only needs to be large enough to enforce international law and order and prevent violent clashes between nations. Lower overall military expenditure and the absence of an arms race may allow the funneling of resources into functions that improve living standards for all. Concepts such as UBI (Universal Basic Income) could become feasible in the absence of an international competition for power.

The quest for a single world-government is as old as human civilization. Some of the early kingdoms in the Middle East, Europe, and Asia were under the illusion that they already controlled most of the world and sought to gain control over the rest. King Cyrus of Persia argued that all nations were entitled to have their own kings as long as they all recognized Cyrus as their "king of kings."[163] Cyrus established a multinational empire with a federal government, but it was far from encompassing the whole world.

Unification of large areas of the world eventually became unstable and collapsed partly due to internal corruption and partly due to the underestimation of external threats. However, those unifications had many advantages. The Roman Empire's unification of the Mediterranean region was seen by many as more favorable than the clashes among neighboring states that ensued in the wake of its collapse.[164]

Prior to the twentieth century, all attempts at world unification were either the desire of one government to take over the world, or the

---

[163] "King of Kings" became the standard ceremonial title of all Persian rulers after Cyrus.

[164] C.T. Davis, "Dante's Vision of History," in *Dante Studies*, No. 93 (1975), Johns Hopkins University.

desire of one religion to convert the entire world and to establish a global theocracy. The latter would be a *de facto* world government under the name of religion. The Roman Catholic church once ruled over western Europe and continues to wield significant political and economic influence around the world. The Islamic caliphate controlled vast territories in Asia, Africa, and Europe through military conquest and religious devotion. A more recent religious order known as the Baha'i faith (formed in the 19th Century) has put forth a detailed plan for a world government, with the central belief that the world should be one country and all mankind should be its citizens.[165]

Any type of world government is unlikely to form, partly because nations who see themselves as superpowers are not inclined to support the formation of a world government over which they would have far less control. Despite this resistance and in the absence of a world government, most countries would still prefer to see all countries abide by certain international norms of conduct. For example, in as early as the fifteenth century, the Spanish philosopher Francisco de Vitoria lectured on laws to protect international commerce. Such laws never had an enforcement mechanism but became general guidelines that many countries voluntarily abided by. The establishment of the League of Nations and the United Nations in the twentieth century were attempts at forming a cooperative form of a federated world government. These institutions proved to be very limited in the scope and effectiveness of what they could achieve.

The materialization of a united world administration has always conflicted with either the short-term domestic priorities or the overreaching ambitions of world powers. The role of domestic priorities could be minimized if every country proceeded to split their executive branch into two independent branches: one in charge of domestic affairs and the other in charge of international affairs and defense. This would partially isolate international relations from each country's volatile and unpredictable domestic politics.

---

[165] *World Order of Bahá'u'lláh*, Shoghi Effendi, 1938.

In most countries, this would mean that the State Department (Foreign Ministry) would become another coequal branch of the government with control over the defense forces. The charter of this international branch would be to defend the country and promote international cooperation. Consequently, the assembly of every country's international branch of government could potentially morph into a cooperative world government with minimal interference with the domestic affairs of each country.

> It will be advantageous if countries split their executive branches.

In order for the entire world to evolve into a single Simorgh, there cannot be an information-sharing barrier. In fact, information sharing barriers may be thought of as natural boundaries that will eventually separate Simorghs from each other. An international consolidated information network without any borders and barriers is unlikely to form, but it is what a single global Simorgh would require.

2. Blocs of Nations

This option is more likely to materialize, since it is in line with the current state of international affairs. In international relations, classifying countries as mutual friends or foes is often based on the ease of information flow between them. Such information can be used for law enforcement, marketing and trade, or for security purposes. During the Cold War, there were two major blocs of nations, referred to as the Eastern Bloc and the Western Bloc that tried not to share information with each other as much as possible. More recently, multiple competing blocs are beginning to emerge. By forming trading blocs and defense treaties, nations benefit from expanded commerce and reduced per capita defense spending. Since these blocs don't encompass the whole world, they need not face the challenge of accommodating vastly diverse cultures and needs. Nations that form a cooperative bloc can facilitate the flow of goods and information across their borders. Thus, each bloc of closely cooperating nations can eventually assume a Simorgh identity of its own (Figure 14-1).

Figure 14-1. Multiple Simorghs may emerge on Earth.

3. Individual Nations

Since Simorgh's nervous system is an advanced information network, the existence of fast, ubiquitous, and seamless networks helps to accelerate Simorgh's formation. There is a higher chance for such networks to work flawlessly in a single country than worldwide or even in blocs of nations. Therefore, an individual nation has a higher likelihood of assuming a Simorgh identity. The transition to Simorgh will be gradual, and once complete, no individual human nor any group of humans will be in control of Simorgh or even aware of its presence or its decision-making processes. When some countries move ahead of others in their ascent to Simorgh, the less advanced nations will not perceive anything different except a high degree of automation, efficiency, and speed that Simorgh countries possess.

4. Large Companies

A potentially smaller entity that can evolve into a Simorgh is a large company. Note, that some companies may employ more people than the entire population of smaller countries. For example, Walmart Corporation had 2.2 million employees in 2018, while the population of the country of Iceland was less than 400,000.[166] A corporation that adopts high levels of automation and AI in its daily operations may have enough human and nonhuman nodes to give it the complexity needed to evolve into a Simorgh. A corporation

> Companies can sidestep the unemployment concerns of the unincorporated human population.

even has an advantage over a nation in that it doesn't have to employ low-skilled, non-specialized personnel that it doesn't need, allowing it to sidestep the political problems of unemployment and underemployment among the unincorporated human population.

All unincorporated humans within the geographical domain of a Simorgh that is nation size or larger, will be considered its microbiome. If smaller Simorgh's form on the size scale of a corporation, many neighboring unincorporated human societies will be independent of Simorgh.

5. A Military Organization

A special case of a large company with a high degree of automation, advanced algorithms, and sophisticated communication networks is a military organization. Its potential Simorgh transition means that, like in all other cases, no individual will comprehend its decision-making process. This is unsettling, especially since any ethical principles that may be programmed into autonomous military algorithms are inherently weak and often contradictory. This is unlike the more coherent principles relevant to other automation algorithms, such as autonomous vehicles.

---

[166] *Fortune*, Global 500, 2019.

War itself is fundamentally immoral, for any reason and in any form. Nevertheless, a military organization needs to freely engage in war, and cause carnage if it so desires. The absolute sanctity of human life becomes violated by military algorithms. A military Simorgh will inevitably consider some human life to be expendable. Such violations occur when individuals are tagged as enemy combatants. The tagging by itself sufficiently justifies the attempt to terminate a person's life. The conditions for tagging remain fluid to allow the redefining of adversaries and allies in different confrontations. With this flexibility, a military Simorgh's self-preservation instinct can give it the potential to break free of allegiance to any nation. It doing so, it may tag all threats to itself as enemy combatants and thereby it may pose an existential threat to all humanity.

It is therefore of paramount importance for humans to cooperatively reduce the sizes of military organizations around the world and hopefully eliminate war as a means of conflict resolution before we lose all control over the matter. Already, the close cooperation of large industrial companies with military organizations, has taken some decision-making power away from individual humans. This has happened because military armament development is based on data from war games, and small skirmishes. The automatic analysis of the data leads to recommendations for upgraded armaments. The acquisition and deployment of such newly developed armament by the military is almost guaranteed because of the influence of the manufacturers over the government (Figure 12-2). This is an early example of how over time humans will relinquish their strategic decision-making power without realizing it[167].

The existing advanced communication networks of a military

---

[167] President Eisenhower's farewell address on January 17, 1961, included a warning about the influence of the "military industrial complex" possibly because he was not able to do anything about it. Furthermore, the companies that form the complex are not to blame. Each board of directors tries to maximize profits as they are expected to. The congress is not the moderating force either, because it takes courage for a representative to oppose the norm. Even absolute rulers in dictatorships may by powerless against the trend, because a dictator cannot oppose his pillars of support. Is the genie of a military Simorgh already out of the bottle?

organization and its rigid command and control regiments make it a fertile ground for a pioneering transition to Simorgh. The urgency of preventing such militaristic transitions cannot be overemphasized.

6. Single or Competing Simorghs

The nature and behavior of a Simorgh will depend profoundly on the presence or absence of competition. A global Simorgh that dominates the Earth without competition doesn't need to strive for efficiency. It can afford to support low-skilled, low-productivity humans if encouraged by its initial programming. Evolution and progress do not necessarily require competition. Therefore, a global Simorgh will continue to evolve, grow, and advance based on the survival and prosperity instincts bestowed to it by its human programmers at an early stage.

In contrast, multiple Simorghs will find themselves competing over access to resources and growth space. From one perspective, this will be a healthy competition because it will promote efficiency and accelerate progress. From another perspective, higher efficiency could mean less reliance on human nodes and less benevolence toward unincorporated humans.

A different reaction to competition that may apply to smaller scale Simorghs is specialization, and growth into niche areas where competition is minimized. For instance, one Simorgh may become focused on energy generation and its distribution to other Simorghs, while another may become in charge of hardware upgrades to multiple Simorghs. This is analogous to the diversification known as ecological specialization that evolves in nature when a population growth faces resistance from another colony[168].

Summary

It is unlikely that we will evolve into a single global Simorgh. It's

---

[168] V. Devictor, et. al., "Defining and measuring ecological specialization," *Journal of Applied Ecology*, 2010, 47, pp 15–25.

much more probable for blocs of nations, individual nations, or even large organizations to assume a Simorgh identity. If the entire world were to evolve into a single global Simorgh, it could potentially be very kind and compassionate to unincorporated humans. Multiple Simorghs will find a competitive advantage in minimizing their costly support for low-skilled humans. This will distance Simorgh from unincorporated individuals. Multiple Simorghs of various sizes and composition may evolve. The only ones that we need to be weary of are military Simorghs, that could potentially become existential threats to all humans.

Chapter 15: Lifespan and Procreation

Living organisms reproduce by either mating or cloning. Cloning, or asexual reproduction, is observed in bacteria, some plants, and fungi. It creates offspring that is genetically identical to the parent.

Reproduction by mating or pairing requires sexual dimorphism and limits population growth compared with asexual replication. In paired reproduction, over half of the population is incapable of bearing offspring, including males and those who do not find mates. This high reproductive cost is obviously an evolutionary success and must be worth the advantages that it offers in disease control, genetic diversity, and by preventing many genetic defects from infecting the offspring. [169]

The evolution of mating started with the fusion or conjugation of cells and the development of gene exchange. These natural processes would transfer parts or all of one cell into another compatible cell. Offspring generation would follow by splitting, or mitosis. Larger cells formed this way had more nutrients essential for the survival of the offspring. This process also had an evolutionary advantage over self-replication because of the gene diversity that it generated. But it created a conflict with respect to the organelles that each cell

---

[169] R. W. Gerard et al., "Chapter 7: Introduction to the Cellular Basis of Inheritance," in Concepts of Biology (Washington, D.C., VA: National Academy of Sciences, 1958).

contributed to the offspring. It either created an overpopulation of organelles, or the reproduction of organelles was curtailed in a less than optimal way. Over time, the conjugating cells evolved into two sexes, with one being small and agile, and the other being large with nutrients and organelles. This pattern migrated to larger organisms and sexed reproduction became the norm. In animals, competition among males for mating privilege, as well as sperm competition, favor the propagation of healthy genes.

Figure 15-1 A pair of birds tending to the young. Typically, the pair stay together for one reproduction cycle until the offspring become independent. (https://animals.desktopnexus. com/wallpaper/514858/)

When humans lived in small groups, mating had to be uninhibited, with the exception of competition among males that led to hierarchical mating privileges. Genetically, chimpanzees and bonobos are the closest primates to humans. Chimpanzees practice both free and hierarchical mating depending on the degree of control exercised by the alpha male. Bonobos on the other hand, are matriarchal, and their interactions are generally peaceful. Also, frequent mating and sex-play are rampant in their daily lives.

Monogamy

Prior to civilization, small groups of humans initially had similar group-oriented mating patterns, but later, monogamy became the norm among city dwellers. There are many theories that attempt to explain this shift.

Monogamy is not very common in nature. One exception is with birds that jointly raise their offspring (Figure 15-1). Joint caring increases the survival chances of the newborn, as one parent can protect the young while the other searches for food. Monogamy in birds typically lasts for one reproductive cycle until the young gain independence. If a similar cooperation urge existed in humans, the duration of monogamy would have to be several years for the young to gain enough independence. The duration could be longer in the case of multiple children. If a couple has been together for that long, they may not have the desire or energy to end their relationship or seek new partners later in life The weakness of this explanation comes from the fact that human couples didn't live in isolation. The extended family always participated in looking after the children.

Another impetus for monogamy may be related to civilization and the growth of human communities that subsequently increased the potential for sexually transmitted diseases. Such diseases would be easier to control in small communities than in large cities. The fear of disease may have been a factor in the development of lasting mating partnerships among humans.

A third possible factor was the development of long-term companionship that was uniquely strengthened among humans with the advent of language. Comfort with each other's companionship as a reason for monogamy would explain yet another behavior: In many human communities there was a higher emphasis on social monogamy than on genetic monogamy. Social monogamy is when a couple decides to stay together despite "external affairs" and promiscuity. At the time of European arrivals, Native Americans were documented as having been monogamous but open to non-monogamous casual relationships. The genetic lineage of the children was not of much concern, as they collectively belonged to the tribe. Similarly, ritual partner swapping was practiced by some Eskimo and Native American tribes despite the underlying social monogamy. Likewise, ritual fertility parties were conducted in various regions of the old-world, including Egypt, Greece, and Rome even though monogamy persisted.[170] We can conclude that social monogamy or

---

[170] Michael Castleman, "Orgies through the Ages," Psychology Today, September 4, 2018.

the need for prolonged companionship supplemented by care for children must have been stronger driving forces toward monogamy than disease avoidance.

Within the last two millennia, monogamy continued to be the standard practice though polygamy also persisted sporadically in some communities. Over time, relationships became regulated by social laws of property ownership, inheritance, marriage, and divorce. Marriage often gained the qualities of a business deal between two families. Precious goods were exchanged in the form of a dowry for example. There were negotiations and written agreements, as in any business contract. A backlash against this in some western societies has placed more emphasis on mutual attraction and a period of engagement prior to marriage. The "success" rate of marriage on the other hand has not changed much, if we define success as a mutually fulfilling and long-lasting relationship[171]. It should be noted that satisfaction in marriage is influenced by the couple's initial expectations of each other and their married life. Such expectations are of course shaped strongly by social norms. Idealistic expectations often lead to disappointment.

Non-conformity

As a consequence of civic and religious laws that regulated reproduction, sexual relationships became narrowly defined as the act needed for procreation, even though it is rarely practiced for that purpose. This strict definition tagged people who didn't conform to it as "abnormal" and strained some relationships by creating performance expectations beyond mutual consent. It has also caused a recent backlash by groups who demand recognition of homosexuality, gender fluidity, etc. Even expanded definitions of a sexual relationship tend to be defined in a traditional male centric way by including a penetration act. A better and more encompassing

---

[171] Since historical statistics are not available, his statement is supported by surveys of traditional vs modern societies: "If you want to be happy for the rest of your life - Study finds women of faith most satisfied in marriage", Catholic News Agency, May 22, 2019.

definition of sexual relationship would be "any intimate and private physical contact that the participants enjoy". This definition diminishes the need for the categorization of sexual preferences.

Additionally, the strict Western definition of marriage as the union of one man and one woman is likely to become relaxed. Polygamy, polyamory, and same-sex marriage are becoming more common. Legal alternatives to the institution of marriage may be created, such as an extended "domestic partnership" that could be the union of any number of participants (Figure 15-2). Of course, there needs to be standard agreements that protect the rights of any resulting offspring (for as long as reproduction requires a human placenta). Furthermore, the rights and responsibilities of each party needs to be well defined in any such agreement, and the terms of expanding or shrinking the partnership need to be specified. Even though the laws of marriage will be generalized and relaxed over time, managing a plural relationship is difficult for the participants, and if history is any guide, most people would prefer to stay away from it and revert to monogamy, even if it is only social and not necessarily genetic monogamy.

Figure 15-2 The 1969 Columbia Pictures comedy-drama film: "Bob & Carol & Ted & Alice" depicts some of the problems in managing plural relationships. From left to right: Elliott Gould, Natalie Wood, Robert Culp, and Dyan Cannon.

Future Developments

DNA sequencing and genetic engineering are likely to revolutionize the approach to reproduction in human societies. Already the availability of Preimplantation Genetic Diagnosis (PGD) has increased the popularity of in-vitro fertilization. PGD prescreens for known genetic disorders and allows sex selection, where permitted by law. In the future, the contributors to the offspring's DNA may not necessarily be just the two parents. Desired genetic traits may be deliberately spliced into the DNA. This may be done for correcting a particular known flaw in the DNA, in which case diseases will be prevented, and the outcome is not very risky. Splicing to change a particular function or characteristic of the body, on the other hand, is not simple. Altering any gene needs to be done in coordination with many other genes in the DNA in order to yield the desirable outcome. Local splicing of the DNA molecule at will, may have unintended consequences due to "gene interaction," which occurs when one gene is changed, causing changes in the expression of other genes. Therefore, any modification or "enhancement" of the DNA needs to be done with full accounting of its effect on all related genes that may also need to be modified accordingly. In other words, only a certain combination of changes will yield favorable results, while others will not be viable.

In the future, it may be possible to simulate the effects of any gene editing ahead of time. This will show all the needed changes for the desired outcome without unwanted side effects. With this capability, humanoids may be synthesized as intended without any surprises. Eventually, some standard and "tried" DNA templates will be identified as the best overall options to be made available "off the shelf".[172] They can be customized slightly according to the preferences of the "parents" or the initiators.

Taking this a step further, the nourishment and growth of the embryo doesn't need to be done in the biological mother's body. The term

---

[172] The path to developing simulation tools or standard DNA templates is not clear. Trial and error in human genetic modification is neither legal nor ethical. Even performing animal experiments ahead of time is inhumane.

"biological mother" will gradually lose its meaning. Every mother may be considered a surrogate mother to a gene-edited embryo. Eventually in the Simorgh society, artificial placentas[173] will separate the reproduction of humans from the sexual relationships among them. Unlike in nature, the cloning of future human species will not make them prone to disease, thanks to Simorgh's advanced internal defense mechanisms that would protect the health of each offspring. So, as discussed earlier, a certain "eusociality" may emerge in the form of a limited number of "safe" and accepted DNA templates to be used in reproduction. Each template would be for a different type of human being or humanoid. Reproduction of each humanoid type will be done on an as needed basis, similar to the growth and replacement of cells within a human body.

> A certain "eusociality" may emerge in the form of a limited number of "safe" and accepted DNA templates to be used in reproduction.

Lifespan and Immortality

Humans may be the only species on Earth who are aware of their own mortality. This can cause anxiety and discomfort for which some people seek refuge in spiritual and religious practices that generally accentuate the immortality of the soul. Aside from spirituality, the desire to delay aging and death has been universal throughout human history. In ancient Greece, people thought that *Ambrosia* was the food that gave gods immortality and could therefore do the same for humans if they gained access to it. The *elixir of life* was coveted everywhere from ancient China to Europe. Gold and mercury were often among the suggested ingredients of such magical potions because gold didn't seem to age. In sixteenth century France, drinking concoctions of gold or its compounds was thought to prevent aging (Figure 15-3).

---

[173] In 2019, Eindhoven University of Technology received a Future and Emerging Technologies grant to develop an artificial womb within the next decade. The immediate goal is the survival of extremely premature babies.

Figure 15-3. Lady Diane de Poitiers, mistress of Henry II of France was one of many nobilities in sixteenth Century France who believed in drinking a gold elixir to delay aging. This practice may have been the main cause of her demise at age 56. [174]

In recent years, a more scientific quest into the causes of aging and death is generating new realistic hope for extended youth and delayed mortality. Already the combination of better hygiene, nutrition, and medical care has doubled human life expectancy within a century. For a long time, the cause of aging was thought to be the inevitable wear and tear of cells and body parts. In the late twentieth century, attention was focused on oxidative damage to cells, and antioxidants were thought to be anti-aging agents. Lately, additional biological processes in the body have been identified that may be responsible for aging. The process of aging is not inherently necessary in living organisms; since every cell in the body is potentially replaceable, the body itself should be able to rejuvenate and doesn't necessarily need to age. Some living organisms such as giant sequoia trees can live for thousands of years.

A study of mortality rates as a function of age in humans (Figure 15-4) led to a thought-provoking conclusion in the late twentieth century. Mortality rate starts at a relatively high level in infants, reaches a minimum in adolescence, and increases steadily after that.[175] First it was noticed that this trend was universal around the

---

[174] Mathilde Thomas, *The French Beauty Solution: Time-Tested Secrets to Look and Feel Beautiful Inside and Out,* Penguin, Jul 14, 2015.

[175] The French demographic situation in 2004 published by INSEE.

world, regardless of race and access to medical care. Later it was fond that many other species even including insects, manifested the same overall behavior scaled differently in time. The conclusion was that the curve represented the effectiveness of sexual reproduction in cleansing defects in the genome.

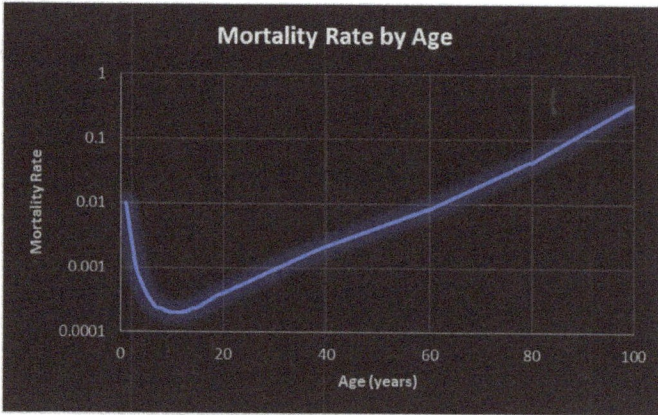

Figure 15-4. Mortality rate as a function of age for US population in 1999. The adolescence accident peak has been removed. The shape of the curve is characteristic of multiplication by sexual reproduction. The minimum mortality rate coincides with the onset of sexual maturity.

Sexual reproduction can only cleanse defects that manifest themselves before the age of reproduction. For example, imagine a gene mutation that causes a person to lose a major bodily function such as eyesight at the age of ten. In the wild, this person would not have much of a chance to propagate his gene to the next generation. If on the other hand the gene mutation causes the loss of eyesight at age fifty, by the time the defective gene manifests itself, it has already been passed down to the offspring. Sexual reproduction does not have a way to cleans the genome of defects that manifest themselves after the age of reproduction. One theory of aging states that such late-acting defects accumulate in the genome and cause degradation of the body as we get older. This effect is included in the following list under the evolutionary aspect of aging along with a few of the other factors that contribute to the aging process.

- Evolutionary

  - Reproduction only cleanses the genome of defects that manifest themselves at a young age.

- Biological

  - Free-Radical damage to lipids, proteins, and the DNA are a natural consequence of oxidative metabolism.

  - Damage from sunlight's ultraviolet radiation is a known contributor to skin aging.

  - Cells age and stop to replicate when their chromosomes run out of protective endings (telomeres) after multiple replications.

- Physiological

  - Misfolding and aggregation of specific proteins cause a variety of neurodegenerative diseases associated with aging, including Alzheimer's disease.

There is a very rare disease by the name of Progeria or more specifically Hutchinson-Gilford Progeria Syndrome (HGPS),[176] which is a disease of premature aging. Children with Progeria don't live much past adolescence. There is a total of a few hundred known cases in the world and there is no cure for it. The disease is caused by mutations in a gene that encodes a useful protein called nuclear *lamin A*, whose function is to protect the cell's nucleus. The mutation leads to the production of a toxic truncated version of this protein, called progerin, that prevents or distorts normal cell division. This protein is abundant in the cells of patients suffering from progeria. It turns out that progerin is also found in small amounts in healthy tissue, and accumulates in the body during normal aging. It is obvious that progerin is a contributor to aging, though it is not the only one. It is expected that any treatment developed for HGPS will also be able to

---

[176] Progeria Research Foundation: progeriaresearch.org.

contribute to the slowing of normal aging in humans.[177]

Active research in causes of aging together with advancing gene splicing and intervention techniques will at least gradually increase the life expectancy of humans and may eventually lead to immortality. Either long life or immortality will force the lowering of birth rates for controlling the population growth. Human reproduction may be triggered only on an as needed basis. Still, the question of whether immortality is a gift or a curse remains unanswered. Some humans or humanoids with long life expectancy may prefer to impose a known time limit on their own lives.

Simorgh's Procreation

Circumstances may arise in which a single Simorgh would benefit from splitting into two or multiple entities. Alternatively, conditions may exist that would entice two or more Simorghs to merge. Finally, one or more Simorghs may decide to spawn a new offspring elsewhere.

For Simorgh, the question of procreation is somewhat different than in the animal world. It is more akin to corporate spinoffs, joint ventures, and mergers. An offspring that inherits selected characteristics of two Simorghs would be analogous to a joint venture created by two companies. It is likely for such a joint venture to receive donated nodes and algorithms from each parent before it can thrive on its own.

Another form of offspring generation that is similar to cloning is a spin-off from a large corporation. In this case, the initial nodes and algorithms will be provided by one parent. the new offspring will have a chance to survive with these inherited capabilities. As the

> Simorgh's procreation is more akin to corporate spinoffs, joint ventures, and mergers.

---

[177] Ergin Beyret, et al, "Single-dose CRISPR–Cas9 therapy extends lifespan of mice with Hutchinson–Gilford progeria syndrome", Nature Medicine, February 2019.

young Simorgh grows, it will have the opportunity to develop new algorithms and new internal architecture that may be distinct from the parent.

As stated above, the reason for paired reproduction in animals is mainly disease prevention. Individuals with genetic defects may not be able to procreate, thereby eliminating the defect from the gene pool. The effect of retroviruses that alter the genetic code may be limited to one individual if the alteration does not replicate into succeeding generations.

Simorgh is also vulnerable to virus infection. Algorithms that run the different body functions of Simorgh are prone to virus attacks instigated by unincorporated humans with mal-intention or a hostile Simorgh. Such an attack may be an attempt to trigger a favorable response from a body part of Simorgh. For example, one can imagine that an energy management module of Simorgh could be tricked by a software virus to allow siphoning off part of the energy for use outside of Simorgh's body.

It is unlikely though that the incentive for offspring generation in Simorgh would be disease prevention. Simorgh's defenses will be advanced enough with the ability to evolve rapidly to eliminate the threat and repair the damage. It is more likely that any offspring generation will be for performing a mission that is different from that of the parent(s). For example, the mission may be extraterrestrial exploration or the harnessing of a new source of energy.

Finally, if one global Simorgh forms and continues to encompass the whole Earth, its extraterrestrial growth may not be in the form of a separate offspring, but rather in the form of a creeping plant that always maintains contact with the original root, even though each section may develop its own root and be potentially self-supporting.

Summary

Aging and sexual reproduction among animals are meant to preserve the species and prevent the propagation of some genetic mutations and viral diseases to successive generations. Simorgh may be immune to such ailments as long as it can safeguard its algorithms from malicious viruses. Simorgh's procreation will be similar to corporate spinoffs and joint ventures. Simorgh will not have built-in mechanisms for aging and mortality. The regenerative capability of each of its organs should protect it against obsolescence.

Chapter 16: SETI (Search for Extraterrestrial Intelligence)

For centuries, most people believed in the very self-gratifying geocentric image of the universe. The sun, moon, and stars revolved around the Earth, where humans reigned supreme over all other life forms. Therefore, it was not unreasonable to think that the whole universe existed just to serve humans. This was the basic assumption in some major religions.

Humans were physically separated from the realm of polytheistic gods who ruled the universe. They lived in faraway places inaccessible to humans. Some gods lived on top of tall mountains and some lived in the heavens. Some prominent thinkers have occasionally considered the possibility that other habitable worlds existed that humans were unaware of. In the writings of some Greek philosophers, medieval thinkers, as well as in the scriptures of Jainism, there are references to a plurality of the worlds.[178] However, the nature of such plurality and whether this included the heavens was not deliberated. It appears that ancient thinkers were visualizing something akin to parallel universes rather than extraterrestrial life. Such discussions were purely philosophical and not cosmological.

---

[178] *Plurality of worlds: The origins of the extraterrestrial life debate from Democritus to Kant* Steven J. Dick. New York: Cambridge University Press. 1982.

Early popular fiction was alive with narratives of travelers to mysterious lands where strange creatures lived. One book titled *True History*, written by Lucian of Samosata around 175 CE, describes travelers who are thrown into the sky by a violent whirlwind and arrive on the moon a few days later. There they find that the inhabitants of the moon are at war with people of the sun.[179] Both sides used strange transport vehicles, and their environment was full of unfamiliar creatures (Figure 16-1). Such stories were read as fairy tales and their underlying assumptions about extraterrestrial life were never taken seriously.

Figure 16-1. In the war between the people of the sun and the people of the moon, Lucian describes the presence of bizarre creatures like space spiders. (*Lucian's True History*, Illustrated by Willian Strang, J. B. Clark, and Aubrey Beardsley. A.H. Bullen, 1894).

The discourse changed after the discovery of the telescope and the erosion of the concept of the geocentric universe. Giordano Bruno, an Italian philosopher of the sixteenth century, argued seriously that every star had its own planetary system that supported inhabitants such as animals.[180] In the nineteenth century, a heated debate began over the existence of intelligent life on Mars. The discussion was fueled by telescope observations of Martian "canals" that were thought to be artificial in nature (Figure 16-2). Astronomers, and most notably Percival Lowell, saw what appeared as networks of straight lines and assumed that they were irrigation canals. The lengthy debate was finally settled in the mid-twentieth century with

[179] Lorraine Boissoneault, "The Intergalactic Battle of Ancient Rome," *Smithsonian Magazine*, December 14, 2016.

[180] G. Bruno, *On the Infinite Universe and Worlds*, Archived from the original in 2014.

better imaging equipment and more accurate observations proving the canals to be mere optical illusions enhanced by wishful thinking.

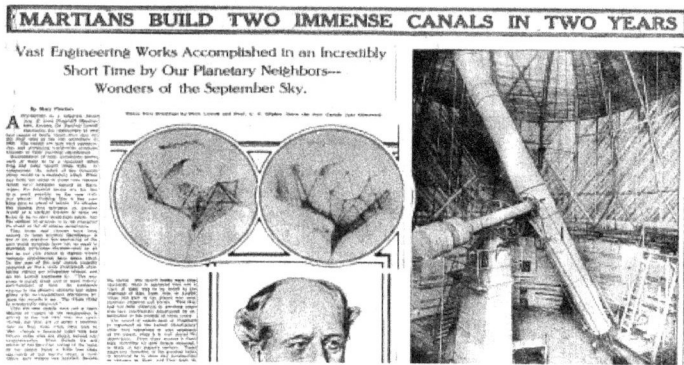

Figure 16-2. *New York Times* article on August 27th, 1911 shows diagrams of canals on Mars and reports the completion of two new canals in a short period of time.

The modern Search for Extraterrestrial Intelligence (or SETI) is an effort to find evidence of advanced lifeforms anywhere beyond Earth. The effort started in the 1960s with the idea of sending or receiving radio signals to distant civilizations. Radio astronomers Cocconi and Morrison[181] showed that a transmitted signal from Earth would be detectable at stars many light years away. For intelligent beings to detect such signals and respond to them would take a very long time because of the distance. If, however, intelligent beings were transmitting such signals to advertise their presence, we could potentially detect them at any time. Large radio telescopes and antenna arrays have been utilized for the purpose of listening to potential messages from outer space (Figure 16-3). However, to date no telltale radio signal has been detected. Most signals that appeared to have deliberate structure were shown to be false positives and were eventually traced back to Earth.

---

[181] Cocconi, G., and Morrison, P. "Searching for Interstellar Communication," Nature, 184, 844 (1959).

Figure 16-3 Left: Five hundred meter aperture spherical radio telescope (FAST) in China is the world's largest, often used for SETI (CGTN July 2020). Right: The SETI Institute's Allen Telescope Array in Hat Creek, California (Seth Shostak, SETI Institute).

There have been a few exceptions over the years with several anomalous, non-recurring bursts of signals that were received but never explained. Further analysis was not possible because they could not be received again. Perhaps the most famous one is the so-called "Wow" signal received on August 15, 1977, by the Big Ear radio telescope at Ohio State University. The signal was a narrowband[182] radio transmission that came from the direction of the constellation Sagittarius. It was tracked for seventy-two seconds. There has never been a satisfactory explanation of its origin, but since no recurrence was detected even with far more sensitive radio telescopes, the one-time event had to be dismissed.

In the year 1960, Freeman Dyson suggested that an advanced civilization may be detectable by their vast energy use.[183] Dyson argued that the alien lifeform may try to harness most of the energy emanating from their local star. Such a star would be abnormally dim because its light would be intercepted for energy use. The term "Dyson sphere" refers to a hypothetical megastructure that would surround a star to capture a large fraction of its radiated energy. If

---

[182] Narrowband means at a precise frequency. There are no known natural sources of narrowband radio emissions. Another type of detected signal known as broadband fast radio bursts (FRBs) last only a few milliseconds and have been detected multiple times. The mechanism of their emission is not known, but they are assumed to have natural causes due to their broadband nature.

[183] F. J. Dyson, "Search for Artificial Stellar Sources of Infrared Radiation," *Science*, 131, 1667 (1960).

a civilization has advanced to that level, we should be able to detect its presence through such signs of massive energy consumption. According to thermodynamics, energy consumption of any kind releases some "waste heat," and this heat should be detectable in the form of long wavelength infrared radiation. Therefore, one signature to look for would be an unusually dim star that is bright in the infrared spectrum. Unfortunately, a similar dimming can be seen when cosmic dust surrounds a star. When the dust is heated by starlight, it gives away infrared radiation, making it difficult to differentiate from a potential Dyson sphere.

To date, neither a Dyson sphere nor any alien megastructure has been positively identified. One case that generated a lot of temporary enthusiasm was the October 2015 discovery of unusual light fluctuations in some images captured by the Kepler Space Telescope. The images were of a star that was later nicknamed "Tabby's Star," in the constellation Cygnus and approximately 1,500 light-years away. The irregular fluctuations sometimes dimmed the light of the star by as much as twenty-two percent and could not be explained by the presence of orbiting exoplanets. No definite explanation was ever found, but an uneven ring of dust circulating the star is one viable hypothesis. Regardless, the search continues. This type of search is known as *artifact* SETI, in contrast with *communication* SETI, mentioned above.

The Future of SETI

In 1964, the Soviet astronomer Nikolai Kardashev devised a three-level scale for measuring a civilization's technological advancement based on its amount of energy use.[184] According to this scale, the three stages of civilization at the cosmic level are:

- Type I or "planetary civilization:" uses most of the energy available on one planet.

- Type II or "stellar civilization:" controls a significant

---

[184] Kardashev, Nikolai, "Transmission of Information by Extraterrestrial Civilizations", *Soviet Astronomy*, 8, 217 (1964).

fraction of the energy output of a star. It may assemble a megastructure such as a Dyson sphere around the star to capture the energy.

- Type III or "galactic civilization:" can harness energy on the scale of an entire galaxy.

We are on the brink of becoming a Type I, or a planetary civilization, by consuming energy at a rate approaching what the Sun provides to the Earth. Our progression toward a planetary civilization is happening concurrently with the emergence of Simorgh on Earth. Our assertion is that any extraterrestrial cosmic level civilization of types I, II, or III that may exist must have first evolved to the complexity level of Simorgh, and the course of expansion from that point on must have been decided by Simorgh. Of course, Kardashev's classification assumes that advanced civilizations will continue to expand in size. We contend that this is most probably not the case, and that a highly advanced Simorgh is more likely to choose not to extend its presence beyond a planet or a solar system (Appendix D).

> Any extraterrestrial cosmic level civilization must have first evolved to the complexity level of Simorgh.

We cannot search distant stars and galaxies for early-stage life, or even for intelligent life that is not highly advanced. The presence of living species at low stages of development won't have a strong enough signature to make them observable from Earth. We can only search for conditions that we think are suitable for life, but we cannot hope for positive identification of life itself. Any quest for early-stage life using our current technology will be limited to our own solar system where we can send probes for investigation.

It can take millions of years for intelligence to fully develop from early-stage life, as it did on Earth. But once complex intelligence appears, its successive stages of advancement accelerate rapidly. It took millions of years for natural processes to create intelligence at the level of a rodent, while it took the new field of electronics only a few decades to create computers that surpassed that. Therefore, it behooves us to focus our attention on searching for highly advanced lifeforms and intelligence. If they exist, they will perhaps be a hybrid form of organic and inorganic beings, similar to humans after merging with technology to form Simorgh.

In the future, the search for extraterrestrial intelligence will become a search by an Earth-originated Simorgh for other Simorgh species in the universe. A search by Simorgh can last much longer than a normal human life. Simorgh can afford to send signals to stars that are many light years away and wait many years to receive a potential response back. It may even have the desire and the required patience to send probes to explore planets outside of the solar system. This is all because Simorgh will either be immortal, or its lifespan will be many times longer than the lifespan of each of its human nodes, even though GM-hybrid nodes of Simorgh may have much longer life expectancy than today's humans. Simorgh could easily live for many centuries and continue to evolve without any mechanism to initiate its demise.

The philosophical importance of SETI is that it can specify a "sphere of solitude" centered at planet Earth within which we know for sure that there is no detectable extraterrestrial advanced intelligence. Currently, the search for early-stage life continues within the solar system, but it is evident that no other intelligent life resides on any of our neighboring planets. Therefore, the radius of our sphere of solitude is from the Earth to our nearest star, Alpha Centauri which is 4.4 light years away.[185] As our sphere of solitude increases in size with further observations, we may someday detect the presence of advanced intelligence in some corner of the universe. Whether making our presence known to such potential beings is wise or not is a question for Simorgh to answer. The decision to contact or not to contact other intelligent life is only one side of the story. Chances are, by the time we discover them, they have already spotted us. If

---

[185] The distance that light travels in 4.4 years. In comparison, light travels the distance from the Earth to our sun in eight minutes.

saturation of growth doesn't happen, and both sides continue to expand to reach cosmic civilization levels II, or even III, it will not be possible to ignore each other's presence.

There is, of course, the possibility that no advanced intelligent life exists anywhere else in the universe, as none has been found in the past fifty years of searching. Despite the optimistic statistical predictions by Drake[186] and others, it may be argued that if the geological history and conditions on Earth were even slightly different, human intelligence would have never emerged. These are conditions such as the presence of the moon to maintain a moderate climate on Earth for a long time; the great oxygenation event to encourage the formation of prokaryotic cells; or the extinction of the dinosaurs to allow the flourishing of mammals, etc. Therefore, even on planets identical to Earth, the possibility of finding high intelligence may be negligible.

A universal saturation of growth, as described in Appendix D, may be a second explanation for the difficulty to locate other advanced civilizations and why it appears that we have not been found by them.

An inherent instability in the development of high technology, may be yet another reason for the dearth of advanced intelligence in the universe. Technology may lead to the self-destruction of civilizations in their adolescence before they can reach the "stellar" level. We may be facing that challenge on Earth soon. Simorgh, if not planned well, may prove to be unstable, and its actions may lead to its own destruction. Another catastrophic scenario would be the armed clash of Simorghs. When such destructive behavior is imminent, humans will be powerless to stop it, because they will not be able to intervene, nor even understand the decision making process of Simorgh. It is incumbent upon our generation to lay the groundworks correctly. We need to remove the option of engaging in armed conflict from all international disagreements.

---

[186] Using Drake's statistical equation, some people suggest that chances of finding advanced intelligence in the universe are relatively high.

Summary

Chances of finding extraterrestrial intelligent life comparable to ours are quite low. Chances will somewhat improve when the search is conducted, with less time constraint, by Simorgh for similar advanced societies in outer space. Our sphere of solitude, within which we know that no other intelligent life exists, has a current radius of 4.4 light-years. This radius may expand in the future with better measurements and observations, or Simorgh may choose to stop all such effort in favor of isolationism. Simorgh may also restrict its own physical expansion beyond the solar system. If extraterrestrial intelligence is found in the future, chances are the finding will be mutual, in which case an eventual contact may be inevitable. On the other hand, all intelligent life may naturally develop instabilities and self-destruct before reaching a stellar level.

Chapter 17: Opportunities and Threats

The emergence of Simorgh will be the most significant development in the history of life on Earth. The fully developed Simorgh will be more magnificent and awe-inspiring than any organic life form, and its path of future development may be incomprehensible to most of us. What is clear is that it will boost the standard of living of its human nodes to new levels of security, comfort, and leisure. It will even bring unprecedented benefits to outsider humans, who can take advantage of the advanced infrastructure of Simorgh. Such infrastructure includes transportation, communication, and service jobs available to unincorporated humans, as there is no spatial separation between them and Simorgh. To unincorporated humans, it will appear that their neighboring corporate buildings have spread their useful infrastructure everywhere.

Opportunities

Algorithms are currently being developed to automate decision making for systems and organizations. These algorithms will eventually include all the instructions for accomplishing various tasks in the organization. Over time, these algorithms will morph into Simorgh's equivalent of the "DNA" code. The algorithms will collectively define the character and instincts of Simorgh. They will also provide Simorgh with a value system to help it make difficult decisions and choose among alternatives. The value system will unavoidably include a strong self-preservation instinct that will guide many of Simorgh's decisions, but it doesn't need to be the only criterion; additional principles such as ethics and moral values can be given to Simorgh that will help it arrive at judgements that are more palatable to humans. Preferential treatment of human nodes over nonhumans, and benevolence toward unincorporated humans, are not traits that Simorgh would naturally acquire, but they may be written in the initial codes, and a Simorgh living in peace will not have any inclination to question or change them.

Currently, ethical decision-making is being advanced for driverless vehicles. Perhaps this will be the first comprehensive platform of ethical values that will be included in a widely used automated system. Driverless vehicles will soon learn to make moral choices that will give them an edge over human drivers. They will be able to choose the least morally objectionable option in an impending accident where bad outcomes are inevitable. Human drivers don't have time to weigh all the options in a split second when something goes wrong and often don't choose the best course of action. Once a comprehensive protocol for fully automated self-driving cars is thoroughly tested and approved, it will serve as the first template for other autonomous machines and for Simorgh's overall ethical framework.

We have the opportunity to lay the groundwork properly for the launching of Simorgh that ensures continued active participation by humans. As AI-controlled systems become more capable, and particularly when they gain the ability to improve themselves, they have the option to exclude humans from their development path in favor of efficiency. But there is also the option of developing enhancements for humans to ensure their continued dominance and value. The goals, instincts, and the value system included in Simorgh's initial programing need to be made dependent on human participation. It is important to emphasize that no one will be programming Simorgh. Its programming is a collection of automation algorithms, each written for a narrow application. We have the opportunity and responsibility to use caution in the initial programming of AI-controlled systems so that human supervision is not excluded, as this will have significant long-term consequences.

Of course, this is easier said than done. Developers of automation algorithms are not aware of the broader implications of their programming. For example, there is no specific algorithm that would decide to replace a human with a robot. Once an employee's job is automated by technological progress, the employee's position becomes unessential, so a cost optimization routine may terminate the human-held position to reduce cost. It is not clear at first glance where in this process we can protect human jobs. The answer is in the need for overarching ethical frameworks that may be able to reverse such decisions. This is similar to a smart car that can learn to place a high priority on saving a human life in an impending accident. The approval process for personnel-related decisions in an organization needs to incorporate ethical values to give a slight preference to human labor when there is a choice to be made. This needs to be followed by offering enhancements and training to human personnel in order to ensure their continued involvement. This is an example of a specific opportunity that we can take advantage of now.

Threats

Armed AI robots programmed for self-survival in the battlefield may be effective war machines but could potentially threaten humans in the long run. There is already a significant development effort toward the realization of such robots. When human-operated military equipment is used, there is always the extra burden of supporting and protecting the operator. Autonomous machines are free from such constraints and may be made smaller and more agile than human-operated ones (Figure 17-1).

Figure 17-1 The MAARS (Modular Advanced Armed Robotic System) has more firepower than suggested by its small size. It can be outfitted with armaments such as machine guns, tear gas, and a grenade launcher.[187]

Lethal Autonomous Weapons, otherwise known as "killer robots," have attracted some public concern and opposition (Figure 17-2). In November 2018, UN Secretary-General called such weapon systems "politically unacceptable and morally repugnant" and urged member states to prohibit them. However, the prohibition of such weapons is pointless, even more so than the prohibition of nuclear, chemical, and biological weapons. The effect of any such prohibition is to drive their development underground. Furthermore, as weapon systems become more automated every day, it is increasingly unlikely that a meaningful and universal definition of "killer robots" can be adopted that every nation would agree to ban. Even if such a prohibition was adopted, its enforcement would at best be possible only in small regional skirmishes. In larger conflicts, no legal prohibition will hold. But the primary reason why such prohibitions are misguided is that they divert attention away from the main problem: the persisting "killer mentality" that governments have retained. Instead of banning certain weapons, we need to strengthen international courts and enforce their judgements in dispute resolution so that people don't resort to killing to settle international differences of opinion.

---

[187] Will Nicol, "The 9 Coolest Military ROBOTS: Maars, DOGO, Etc.," Digital Trends, July 17, 2020.

Figure 17-2. Campaign to Stop Killer Robots, Canada 2018.

Most opposition to autonomous weapons comes from the fear of their potential misjudgment in the field or their built-in biases. However, the far bigger danger is the possible emergence of an autonomous army that we refer to as a military Simorgh. Keep in mind that an autonomous army doesn't mean a lack of human participation; it only means that humans will be taking care of specific tasks and will not be cognizant of any strategic or tactical decision options even

> Weapons prohibitions are misguided because they divert attention from the persisting "killer mentality" that governments have retained.

when they are an integral part of it. Those decisions are made in sequential small steps that no one objects to, as is currently practiced in many sectors of society where everyone focuses on solving tactical problems even when the overall strategy is flawed. A highly intelligent military Simorgh that sees some human life as expendable could potentially break free of all allegiances and, on its own, assume the roles of judge, jury, and executioner.

Another threat, mainly to the GM-humans who are nodes in Simorgh's body, is the potential failure of Simorgh. When a human body dies, all cells within the body die with it. Similarly, if a superorganism fails and dies for any reason, its population of human nodes will also be wiped out. The reason is that after years of having everything delivered to them automatically, human nodes gradually lose their ability to support their own basic needs. In this way, their livelihood will depend on Simorgh's support network, delivery systems, and defense infrastructures. Simorgh's demise will terminate the lives of its human nodes and will initiate the scavenging of its parts.

The death of a human body includes the following steps: brain cells die a few minutes after the pumping of blood stops. Other cells follow, with skin and bone cells being the last ones to perish within several days.[188] Body cells die after the death of their host because they are not capable of feeding and protecting themselves without the infrastructure that supports them. The scavenging of the body starts first from within by the bacteria that live in the body, and later by insects and their larvae. If the body is outdoors, larger scavengers, such as animals and birds, start to feed on it almost immediately.

Using the above analogy, the process of death in a Simorgh starts when food and energy delivery systems cease to function. This will cause the demise of the human nodes and the gradual powering down of electronic nodes. Unincorporated humans will notice the loss of security barriers and will start to scavenge the parts. If other Simorghs exist, they will also take part in the scavenging until all body parts of the dying Simorgh have been claimed or dispersed. These body parts include buildings, computers, robots, and larger entities such as transportation systems and its food and energy supply networks.

We have not speculated on why a Simorgh might fail and die. Internal inconsistencies may develop over time, or unincorporated humans many overcome its defenses and manage to destroy it. Regardless of the reason, the situation would be most dire if only one global Simorgh evolves and dies. Such failure could be a major devastation, as well as a large evolutionary setback.

[188] Tom Scheve, "How Body Farms Work," HowStuffWorks Science, January 27, 2020.

Conclusion

People's response to any potential change in living conditions varies from total resistance to full acceptance. Resistance is the more normal reaction due to our proclivity to stay with the familiar; but this has limitations: It is not possible to stop the changing world, and old lifestyles may no longer fit into the transformed environment. On the other hand, complete embracing of change, often fueled by premature exuberance, has the potential for serious unintended consequences. In most cases, the best choice is to avoid both extremes and manage the change wisely. We are at the dawn of a new era in human living conditions that offers some of us the opportunity to merge with our technology and build a more comfortable existence while paving the way for the emergence of a much more advanced life-form the likes of which has never been seen before. It is incumbent upon us to intelligently navigate this major shift in human society.

The rise of Simorgh is the culmination of human ingenuity and the next major evolutionary advance on planet Earth. It deserves to be viewed more with admiration and engagement than with alarm and revolt. It is unlikely that progress in this direction can be stopped in favor of maintaining the status quo or in favor of retreating to past human individualism. Instead, there is the option for many individuals whether or not to assimilate with Simorgh. In fact, a significant portion of the human population will not join this trend either by choice or by fate. Simorgh will coexist with unincorporated humans and other species on Earth, yet none will comprehend the nature of its existence. Its presence, however, will inevitably impact all life on Earth.

Humans who become incorporated as body nodes of Simorgh will rapidly evolve both by genetic enhancement and further integration with technology. They will become genetically modified human-technology hybrids (GM hybrids). In this way, multiple breeds of humans will emerge that will be different in appearance, function, and capabilities from each other and from unincorporated humans. Unincorporated, unaltered humans will still be able to enjoy a good lifestyle, and benefit from Simorgh's productivity, as long as they are not perceived as a threat to Simorgh.

The status of human nodes of Simorgh versus unincorporated humans is analogous to that of human body cells versus its bacteria and microbiome. Human cells enjoy the security of being provided for by the body's food distribution network and being protected by its very effective defense system. This comfort comes at the expense of having no individual freedom to select among job functions and responsibilities.[189] In contrast, the bacteria that live in and on the human body outnumber the body cells and are not regulated by the human DNA code. They retain their individual freedom, and at the same time, they use some of the human body 's resources in exchange for performing certain services, but they are not necessarily taken care of by the body.[190]

Simorgh doesn't naturally possess a survival instinct nor long-term priorities. This instinct, and other fundamental priorities need to be provided by humans who write the initial algorithms that form the skeleton of its future life code. This will allow us to intertwine the

---

[189] In the sociological terminology of structure versus agency, the behavior of human nodes will be heavily dominated by structure.

[190] It is intriguing to think that a select group of humans will find themselves "raptured" into becoming nodes in Simorgh's body where they will be fully taken care of, while others will be left to earn a living on their own.

survival of Simorgh with the survival of humans and to prevent any competition with humans for survival. If multiple Simorghs emerge on Earth, competition among them may force them to rewrite the codes and squeeze out many humans in favor of efficiency. In contrast, if a global Simorgh emerges, its initial programming can be more favorable to human welfare, and it will have no incentive to rewrite the code. On the other hand, lack of competition in the second scenario may limit Simorgh's healthy evolution and may make it less fit for long-term survival.

Life has survived on Earth because the goal of life is to survive, and any life-form that may have emerged in the past without this goal had no choice but to perish. Simorgh will have a good chance of survival because we will embed the desire to survive in its programming and its structure. However, the path to survival is not unique, as attested by the diversity of successful organisms on Earth. Continued involvement of human GM-hybrids in Simorgh's body may not be necessary, but it won't pose any threat to its long-term survival either. In fact, not only can Simorgh's reliance on its human nodes be programmed, but a degree of benevolence and welfare towards unincorporated humans can also be included in its ethical framework.

We are not capable of fully engineering the behavior of the emerging Simorgh because its future decisions will be on a scale that is beyond our imagination or control. However, at the present time humans are still holding on to the last remnants of their commanding position and can influence and manage the transition properly. In fact, it is the duty of the current human generation to plan, program, and nourish the evolution of Simorgh. We should lay the groundwork for Simorgh's success. At the same time, we need to carve out favorable ranks and roles for humans to hold in future decades and centuries; as Simorgh becomes the dominant living organism on Earth.

The emergence of one or more military Simorghs is of particular concern for human welfare, and unfortunately appears to be highly likely with continued international arms buildup. In a military

Simorgh, the termination of human life is acceptable by design as long as the victim is tagged as an enemy. Since we cannot comprehend nor control Simorgh's decision-making processes, including the identification of friends and foes, a military Simorgh can pose a survival threat to all humans. The continued international arms race can be the social instability that will eventually end human civilization on Earth, not necessarily by a catastrophic clash as thought in the past, but by the loss of human control over institutions of war. This can happen concurrently with passive human involvement at all levels.

Finally, it should be emphasized that we are not capable of consciously programming any aspect of Simorgh. We are programming automation and AI that will become parts of Simorgh's complex code in ways that are not controlled by us. We can only build ethics and human protection in the automation routines and hope that they will shape Simorgh's long term instincts.

Appendix A: The Legend of Simorgh

Simorgh is a mythical bird that appears in various forms in ancient Iranian narratives and in other countries with links to old Persia. In pre-Islamic Zoroastrian texts Simorgh is mentioned as a huge bird nesting on a tree in the middle of the sea. The tree has a potent all-healing medicine and carries the seeds of all plants. There are even earlier references to this legendary bird that date to the Indo-Iranian era of the second millennium BC.

Descriptions of Simorgh have changed over the ages, but its underlying attributes of beauty, benevolence, immense size, and supernatural wisdom have persisted. The collection of old legends known as The Epic of Kings (Ferdowsi, 1010AD) contains episodes that prominently feature Simorgh as a female bird living at the crest of a very tall mountain named Alborz.

Simorgh in "The Epic of Kings"

In one episode that criticizes the social stigma associated with being an albino, Simorgh spots an abandoned albino newborn (later named Zaal). She takes pity on the child and decides to adopt and raise him in her own nest. The child happened to be a prince whose father considered his albino appearance to be a disgrace and ordered the boy to be taken far away from town and abandoned at the outskirts of mount Alborz.

Years later the boy grows to be a very agile and athletic young man who spends most of his time exploring the heights and peaks near mount Alborz. Informants spot him and alert his father to the presence of an unusual young athletic wild man with white hair. His father suspects that this may be his long-lost albino son, and seeks his reunion, but finds the terrain too foreboding for his troops to gain access to Zaal. Simorgh learns about the father's desire to reunite with his son, and breaks the news to Zaal who is initially reluctant to go back, because Simorgh has not only been a caring mother to him but also his teacher and his mentor. Zaal finally agrees to join his father after Simorgh reassures him of her continued support and protection. She gives Zaal one of her feathers to keep and promises to rush to his aid once the feather is set on fire.

In his father's court, Zaal becomes a well-respected and successful warrior. He falls in love with and marries a princess of a hostile lineage. Later their son, Rostam, becomes even more of a celebrated national champion and an undefeated war hero.

Another episode, in which Simorgh plays a key role, relates to the invulnerable crown prince Esfandiar who is vying for his father's throne and insists that the king retire early. There are different narratives of how Esfandiar acquired his invulnerability, and why the protection did not extend to his eyes; possibly because they were closed during his childhood immersion in divine water.

The King, who does not want to reject his son's demand, but is not eager to abdicate in his favor either, repeatedly makes the power transition conditional on Esfandiar 's success in national security challenges requiring body strength and prowess in warfare. In every case Esfandiar returns victorious and proud. So, in his final challenge, the king gives him the daunting task of capturing the national hero Rostam (Zaal's son) and carrying him back to King's court in chains. The fabricated rationale was that Rostam had become too arrogant and a rogue element. Rostam's refusal to be chained results in a vicious and taxing single combat duel. Rostam is heavily injured by the end of the first day of contest, while Esfandiar's invulnerability has kept him fully shielded. The exhausted Rostam goes to his father Zaal to rest overnight and to confer with him about his frustration with Esfandiar. With a sense of desperation, Zaal seeks help from Simorgh by setting her feather on fire. Simorgh arrives to help, offers remedy to instantly heel Rostam's wounds, and tells the father and son about the vulnerability of Esfandiar's eyes. The only way to defeat him is to make a bifurcated arrow and target both of his eyes at once, says Simorgh. The next day Rostam goes to battle well prepared and executes Simorgh's orders flawlessly. Esfandiar is vanquished and the nation loses a highly capable though belligerent crown prince (Figure A-1). Simorgh's intelligence, knowledge, and magical healing powers are on display again in this tragic episode.

Figure A-1. Great warrior Rostam conversing with the dying prince Esfandiar. Tehran City Theatre, Photo by Marzieh Mousavi IRNA

Simorgh in "The Conference of the Birds"

A very different reference to Simorgh is found in the 12th Century book of poetry "The Conference of the Birds" by Attar of Nishapur. This narrative is more philosophical in nature and is a symbolic presentation of a core belief of Suffism.

Figure A-2. The wise hoopoe (bird of Solomon) volunteered to lead the journey. (Birdeden.com)

The story is about a nation of birds that yearns for a worthy leader. The birds gather in a conference to deliberate the matter. The wisest bird among them, the hoopoe (Figure A-2), suggests asking the exalted Simorgh to fill the leadership role. All birds approve, and start to plan a journey to meet with their chosen leader. Simorgh is universally admired and loved by all birds, but she lives on top of a tall and remote mountain that is difficult to reach. The hoopoe volunteers to lead the journey. He warns the birds of the extreme length of the voyage and the seven valleys of hardship that they need to cross. A large number of birds bow out and decide to stay behind after hearing about the challenges. Many more birds either perish along the way, abandon the years-long journey, or choose not to face the challenges of crossing each valley of hardship. Very few birds endure the difficulties and manage to reach the destination where Simorgh is believed to reside. There they find the abode to be devoid of Simorgh. Instead, the stage appears to them as a mirror in which they see themselves, the thirty remaining birds. They realize at this moment that they themselves are collectively the Simorgh that they sought for years. This realization is a play on the word "Simorgh" that may be split into two words: Si and Morgh that mean "30" and "bird" respectively in the Persian language. Therefore, Si Morgh or thirty birds find themselves as having become the legendary Simorgh who was their original obsession and hero.

In Sufism, (as well as in Manicheanism, and Gnosticism) there is a profound conviction that in every human being there is divine spark, although most people are not conscious of it, and that it is possible to attain the divine by suppressing worldly and materialistic aspects of one's existence. The long and difficult journey which Attar represents as the seven valleys of hardship is meant to accomplish exactly that.

The first valley is Quest where you set aside all your beliefs and make your mind open and receptive.

The second valley is love that you need to place above reason.

The third valley is knowledge where you shed all your prior knowledge.

The fourth valley is detachment where you separate yourself from the world.

The fifth valley is unity where you and the universe become one.

The sixth valley is wonderment where you stand in awe and admiration for the beauty of the Beloved.

The seventh valley is Poverty and Annihilation, where you abandon all your remaining earthly possessions and enter the realm of the divine.

The thirty remaining birds who successfully completed all the needed self-cleansing tasks ultimately discover that they had been in possession of their lofty goal all along.

The legend of Simorgh is symbolically in line with the main tenet of this book in that a select group of people with the necessary skills and drive will step up to the challenge of supporting the Earth's next major evolutionary advancement.

Appendix B: Large Numbers

We can easily form a mental image of small quantities of objects and relate to them in our daily lives. For example, if we see one or two swans on a lake, we can retain a mental image of them and remember their quantity the next day. If the number of swans is 5 or 6, the accuracy of our recollection goes down, but we can still count them in our memory. When the number goes beyond ten it starts to strain our ability to recall the quantity form our mental image. If there are more than twenty swans, we simply recall them as "many". The fact that we cannot retain a mental image of larger numbers prompted our ancestors to map the quantity of objects onto tally marks that later evolved into our numbering system.

In our daily lives we routinely encounter numbers much larger than twenty and we feel comfortable with them because we either place them in a small number of subgroups or we relate them to other much smaller quantities. If we see a large crowd of people on the street, we may gauge their number by the much smaller number of street blocks that they occupy. Or we may gauge a large sum of money with the number of houses that it can buy. Without subgrouping or referencing to smaller quantities, we don't comprehend numbers like a thousand or a million. Even with some mathematical training, we are often surprised by the properties of large numbers.

Large Populations

The behavior of a large population is often significantly different from the behavior of its individual members. This can be seen in economics, sociology and natural sciences. As a simple example consider the percentage of a country's eligible population that will participate in an election. Since each citizen can either choose to cast a vote or not, one may think that the participation percentage can range anywhere from zero to one hundred percent. However, a statistician may predict based on previous observations that fifty to sixty percent of the population will vote. This prediction will come true regardless of each individual's freedom to decide. The accuracy of the prediction is higher for larger populations.

Some laws of physics are based on the same concept. Take for instance the air molecules that move rapidly in random directions in a room. At any given time, each molecule may be in one side of the room or the other. It is conceptually possible that all air molecules may momentarily end up in one side of the room, temporarily depriving the other side of all air. If there were only one air molecule in the room, this would happen fifty percent of the time. With two molecules the probability drops to twenty five percent and so on. With the very large number of molecules in the room the probability is so small that the laws of gases in physics declare it impossible. The gas will always fill the available space.

In general, it is important to keep in mind that the behavior of a large population is a lot more constrained than its large degrees of freedom would suggest. A noteworthy exception to this rule is when a large interacting population gains consciousness. The collective behavior of the large number of cells in the human brain cannot be predicted by statistical analysis.

Calculation with Large Numbers

Consider two very large prime numbers. They can be multiplied together easily, but the reverse operation of prime factorization is computationally foreboding,[191] as even the most efficient algorithms take a long time to perform the task. As an example, take a simple small number like 323. It has prime factors 17 and 19. Meaning that 323 = 17×19. This prime factorization can be done by pencil and paper in a few minutes. Now consider a larger number, like the following twenty-four-digit number with its prime factors: 622,851,271,749,310,350,329,267 = 974,892,354,701 × 638,892,354,367. It is very difficult to find the prime factors for this number manually, but it *can* be factored relatively quickly with a computer algorithm. However, as the number of digits goes up, the required computation time increases exponentially. Public key cryptography takes advantage of this property of large numbers to devise methods of encryption with a public key, while decryption would require a private key.[192] This way, the interception of a new public key is not of much help to the decoder. With current available computing power, a code that consists of a few hundreds of decimal digits may be considered "unbreakable."

---

[191] Example: RSA encryption system 1978.

[192] R. L. Rivest, A. Shamir, and L. Adleman, "A Method for Obtaining Digital Signatures and Public-Key Cryptosystems," Communications of the ACM 21, no. 2 (February 1978): pp. 120-126.

Appendix C: Scales and Layers of Reality

Our mental image of the world is an abstraction by our mind, meant to facilitate our interaction with the objects around us. It is shaped by information we obtain through probing the environment. Our senses, dominated by vision, conduct incessant probing of our immediate neighborhood. The way we see objects is by probing them with a very narrow portion of the electromagnetic spectrum, called visible light, and we sense the reflected light with our eyes. Similarly, our sense of smell probes chemicals in the air, and our sense of touch probes the rigidity, texture and temperature of an object. Though the signals generated by each sensory organ are correlated with particular properties of the external world, the sensations that they trigger in our brain are completely arbitrary and subjective.

The shapes, colors, textures, smells, and sounds that we sense are unique to each of us and tailored to our needs, but they are not absolute realities. They change not only from person to person and species to species, but also depend on the probing method and the tools we use. A bat's world has to appear significantly different from ours because it uses ultrasonic probing instead of optical sensing. We don't even have a reason to believe that two people's perceptions of shape, color, or sound are the same. We just learn to give them the same names for the purpose of conversation. Microscopes and telescopes open our eyes to worlds that are significantly smaller and larger than the scale of our daily routine. These worlds are unlike anything we see with our unaided eyes, not only in appearance but also in function. We do not recognize our own skin under the microscope for example, and we can see microscopic organisms in what our unaided eyes show us as clear water. What we perceive as the true nature of the world is both probe-dependent and scale-dependent.

Layers of Time

When we watch grains of sand falling in an hourglass or water flowing in a river, we think we can feel the "passage" of time. The concept of time is very familiar to us. We think we understand it, and in most cases, we even complain that we don't seem to have enough of it. However, the reality of time is profoundly scale dependent. Investigating time on a microscopic scale gives us a surprisingly different image. If we record the motion of molecules and particles under a microscope, we find it impossible to tell if the video is running forward or backward. Microcsopic motion is time reversible. Laws that govern the movement of small particles work equally well under time reversal. Time has no forward or backward direction. This is not the case in the macroscopic world. If we record any of the macroscopic processes of our daily lives, we find it easy to tell which way the video needs to run. Watching spilled water recollect back into the glass, we clearly see an impossible event that identifies the backward running video. The spilled water is more "disordered" than water in the glass, and the "arrow of time" always points in the direction of increased disorder. Disorder is another statistical concept that applies to macroscopic systems. The state of disorder is called "entropy" in thermodynamic terminology. The arrow of time is in the direction of increased entropy. Microscopically, most events can be reversed, so time has no direction. Macroscopically many events cannot be reversed, and time has the meaning that we are familiar with. We see that the reality of time and its directional nature are scale dependent.

Appendix D: Saturation of Growth

Both Kardashev's classifications of advanced extraterrestrial civilizations, and Dyson's argument for detecting such civilizations by their energy use, are based on the assumption that civilizations will continue to grow and their appetite for energy-use will increase with time (Chapter 16). This assumption may have to be re-examined to answer the question of whether a civilization's continued growth is a fundamental tendency.

Population growth among living cells, wild species, and early-stage civilizations has been a mechanism for survival, as they find safety in numbers. Such growths increase the chances of endurance of the species in the presence of threats like food shortages and external attacks. Population growth will occur whenever possible, though the rate can vary. Natural population growth is either exponential when nutrients and space are abundant, or logistic, when limited resources are encountered and population growth saturates. Human population remained relatively stable with birth rates and mortality rates in balance for centuries but started to rise in the nineteenth century and exploded in the twentieth century, primarily due to improved healthcare and better nutrition. Some thinkers went as far as suggesting that human population growth could only remain in check by either famine or war.[193] In recent decades this dire prediction

---

[193] Thomas Malthus "*Principles of Political Economy*" (1820).

has proven to be not the only option.

The seriousness of overpopulation threat has been diminished by recent demographic shifts in human societies. People have been transitioning from agricultural to urban living where children are seen more as financial liabilities than helping hands on the farm. Consequently, fertility rates have dropped, and population growth has slowed without a major war or a widespread famine. According to some projections, the human population will reach a peak of around ten billion before reversing and possibly stabilizing in the vicinity of seven billion. (Figure D-1).

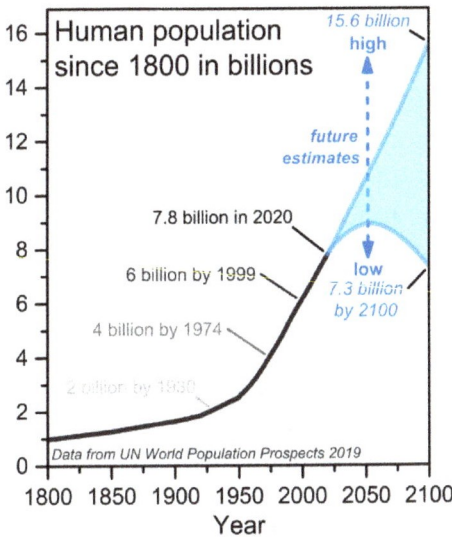

Figure D-1. Depending on assumed fertility rates, human population can continue to grow, or it can reach its peak and reverse direction (https://population.un.org/wpp).

It is not clear if an advanced civilization like Simorgh will possess an urge to continue to grow its human population. Fertility rates of human nodes can be managed in order to maintain the desired population level. Simorgh does not need to continue expanding its domain of influence in the absence of resource competition or any threat to its survival. Its survival instinct would perhaps push it to colonize a few other planets and space stations in the solar system in case the Earth faces a threat such as an asteroid collision or an extreme climate change that it cannot prevent. But it may not find a good reason to expand beyond the solar system. When each Simorgh node is capable of experiencing any environment through virtual reality, why would it be inclined to establish a physical presence in distant locations?

Simorgh's need for energy could also reach a saturation level. Energy consumption per household is in decline in most industrialized regions of the world (Figure D-2). The obsolescence of incandescent lights has dramatically increased residential and industrial lighting efficiency. Energy efficient homes use better insulation and passive solar designs to reduce energy consumption for heating and cooling. With the continuation of this trend, our hunger for energy is likely to reach a limit and possibly decline when population growth reverses direction.

U.S. Residential Energy Consumption

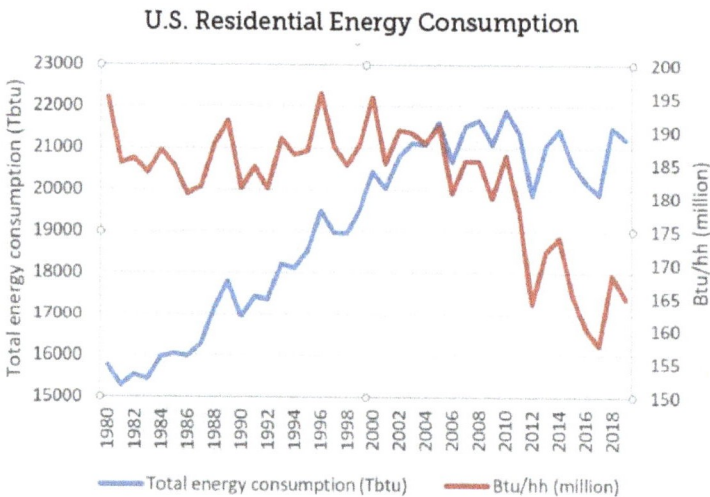

Figure D-2. Residential energy consumption in the U.S. has plateaued despite rising population (blue curve). Energy consumption per household has been decreasing since the early 2000's. (ACEEE: American Council for an Energy-Efficient Economy).

It is plausible to think that Simorgh will need an ever increasing computational power to continue to improve its internal operations and enhance its own life experience as well as the life experience of its human nodes. Increased computational power requires additional energy. In 2020, data centers for the internet consumed about one percent of the world's total electricity. Google's energy consumption grew by a factor of 3.7 between the years of 2011 and 2018[194]. This trend, however, doesn't need to continue indefinitely.

Similar to population rise that will plateau by lowering the reproduction rate, the rise in computational power consumption will be held in check by improved computing efficiency. A measure of computing efficiency is flops per watt (floating point operations per watt of energy). This measure has been exponentially increasing since the advent of electronic computing. Highly efficient computers achieved over 2 Gflops/watt (two billion flops) in the year 2020, which is an improvement by nearly a factor of ten in the span of only five years. Can efficiency increase to the level that computation can be done with almost no energy consumption? The answer is no, but the lower limit is not known yet. There is a projection known as Landauer's limit that places the maximum achievable efficiency at ten million times the current level or twenty million Gflop/watt. This limit may be reached in the not-too-distant future. If efficiency continues to improve by a factor of ten every five years, Landauer's limit may be reached in thirty-five years. Landauer's limit is based on entropy loss due to the erasing of information that occurs in computing. Energy consumption is needed to prevent a net entropy loss that is forbidden in thermodynamics. Alternative types of computing such as "reversible computing" are under development that may not be bound by this limit. Even in the absence of such novel techniques, attaining Landauer's limit itself would provide Simorgh with significant computing power with very low energy consumption.

The growth of advanced civilizations in both size and energy consumption may eventually come to a halt. There may not be an incentive for continued colonization of extra space, and no need to continue the quest for new energy resources. Simorgh may live harmoniously with nature and its continued progress may only be

---

[194] "Energy Consumption of Alphabet (Google) from Financial Year 2011 to 2019," Statista, March 24, 2021.

in the optimization of its internal operations. The "saturation of growth" hypothesis may provide a plausible explanation for why the Artifact SETI effort has yet to find any extraterrestrial intelligence and may never find one.

Bibliography

"Abraham Lincoln Was a Woman." *Weekly World News*, January 1, 2002.

Adams, Tim. "Nick Bostrom: 'We Are like Small Children Playing with a Bomb.'" The Guardian. Guardian News and Media, June 12, 2016. https://www.theguardian.com/technology/2016/jun/12/nick-bostrom-artificial-intelligence-machine.

Agarwal, Nitin. "The Idea of Augmented Reality in App Innovation." Wildnet Technologies, April 30, 2019. https://www.wildnettechnologies.com/the-idea-of-augmented-reality-in-app-innovation/.

Allen, Daniel. "The Rapidly Developing Field of Neuroprosthetics." MedicalExpo e-Magazine, July 7, 2021. https://emag.medicalexpo.com/the-rapidly-developing-field-of-neuroprosthetics/#:~:text=With%20brain%2Dmachine%20interface%20technology,the%20potential%20to%20work%20miracles.&text=%E2%80%9CThis%20is%20an%20exciting%20time,in%20technology%20driving%20growing%20adoption.

"Archimedes." Encyclopædia Britannica. Encyclopædia Britannica, inc. Accessed August 19, 2021. https://www.britannica.com/biography/Archimedes.

Attār Farīd-ad-Dīn, and Wolpé Sholeh. *The Conference of the Birds.* New York: W.W. Norton & Company, 2017.

Bekhrad, Joobin. "The Book of Kings: The Book That Defines Iranians." BBC Culture. BBC, August 16, 2018. https://www.bbc.com/culture/article/20180810-the-book-of-kings-the-book-that-defines-iranians.

Beyret, Ergin, Hsin-Kai Liao, Mako Yamamoto, Reyna Hernandez-Benitez, Yunpeng Fu, Galina Erikson, Pradeep Reddy, and Juan Carlos Izpisua Belmonte. "Single-Dose Crispr–cas9 Therapy Extends Lifespan of Mice With Hutchinson–Gilford Progeria Syndrome." *Nature Medicine* 25, no. 3 (2019): 419–22. https://doi.org/10.1038/s41591-019-0343-4.

bibliothekia. "Working with Rich Metadata in the Cloud. Are We Ready, or Are We Getting Indigestion?" Cloud Librarian DownUnder, July 21, 2016. https://cloudlibrariandownunder.wordpress.com/2016/07/18/working-with-rich-metadata-in-the-cloud-are-we-ready-or-are-we-getting-indigestion/.

Bithas, Kostas, and Panos Kalimeris. "A Brief History of Energy Use in Human Societies." Essay. In *Revisiting the ENERGY-DEVELOPMENT Link: Evidence from the 20th Century for Knowledge-Based and Developing Economies.* Switzerland: Springer, 2016.

Boissoneault, Lorraine. "The Intergalactic Battle of Ancient Rome." Smithsonian.com. Smithsonian Institution, December 14, 2016. https://www.smithsonianmag.com/history/intergalactic-battle-ancient-rome-180961416/.

Bruno, Giordano. *On the Infinite Universe and Worlds*, 1584.

Buchanan, Bill. "IBM's Greatest Challenge?" Medium, October 12, 2020. https://medium.com/asecuritysite-when-bob-met-alice/ibms-greatest-challenge-f03bc6830fa1.

Buis, Juan. "This Girl Isn't Real, and It's Proof That CGI Isn't Creepy Anymore." TNW | Creativity, April 27, 2021. https://thenextweb.com/news/cgi-girl-isnt-real?utm_campaign=share%2Bbutton&utm_

content=This+girl+isn%27t+real%2C+and+it%E2%80%99
s+proof+that+CGI+isn%27t+creepy+anymore&utm_
medium=referral&utm_source=copypaste.

Busato, Gianluca. "Why You Should Start Using Augmented Reality (Ar) and Gamification?" Medium, March 27, 2018. https:// medium.com/enkronos/why-you-should-start-using-augmented-reality-ar-and-gamification-1fdb1e1e57e6.

"Cashiers : Occupational Outlook Handbook." U.S. Bureau of Labor Statistics. U.S. Bureau of Labor Statistics, April 9, 2021. http:// www.bls.gov/ooh/sales/cashiers.htm.

Castleman, Michael. "Orgies through the Ages." Psychology Today, September 4, 2018. https://www.psychologytoday.com/us/ blog/all-about-sex/201809/orgies-through-the-ages.

Cna. "'If You Want to Be Happy for the Rest of Your Life' - Study Finds Women of Faith Most Satisfied in Marriage." Catholic News Agency. Catholic News Agency, May 21, 2019. https:// www.catholicnewsagency.com/news/41343/if-you-want-to-be-happy-for-the-rest-of-your-life-study-finds-women-of-faith-most-satisfied-in-marriage.

Cocconi, Giuseppe, and Philip Morrison. "Searching for Interstellar Communications." *Nature* 184, no. 4690 (1959): 844–46. https://doi.org/10.1038/184844a0.

Collera, Virginia. "Birds Looking for Answers." The Dispenser, January 1, 1970. https://eldispensador.blogspot.com/2012/08/ coloquios-flautinos-pajaros-en-busca-de.html.

Crew, Bec. "Here's How Many Cells in Your Body Aren't Actually Human." ScienceAlert, April 11, 2018. https://www. sciencealert.com/how-many-bacteria-cells-outnumber-human-cells-microbiome-science.

Davies, Ernest Albert John. "Canals and Inland Waterways." Encyclopædia Britannica. Accessed July 27, 2021. https:// www.britannica.com/technology/canal-waterway.

Davis, C T. "Dante's Vision of History." *Dante Studies* 93 (1975).

Devictor, Vincent, Joanne Clavel, Romain Julliard, Sébastien Lavergne, David Mouillot, Wilfried Thuiller, Patrick Venail, Sébastien Villéger, and Nicolas Mouquet. "Defining and Measuring Ecological Specialization." *Journal of Applied Ecology* 47, no. 1 (2010): 15–25. https://doi.org/10.1111/j.1365-2664.2009.01744.x.

Dick, Steven J. *Plurality of Worlds: The Origins of the Extraterrestrial Life Debate From Democritus to Kant.* Cambridge: Cambridge University Press, 1984.

Dyson, F. J. "Search for Artificial Stellar Sources of Infrared Radiation." *Science* 131, no. 3414 (June 3, 1960): 1667–68. https://doi.org/10.1126/science.131.3414.1667.

EETimes. "Digital Data Storage Is Undergoing Mind-Boggling Growth." EETimes, September 14, 2016. https://www.eetimes.com/digital-data-storage-is-undergoing-mind-boggling-growth/.

Effendi, Shogi. *The World Order of bahá'u'lláh*, 1938.

Einstein, Albert, and Alan Harris. *The World as I See It.* New York: Philosophical Library, 1949.

Engstrom, David Freeman, Daniel E Ho, Catherine M Sharkey, and Mariano-Florentino Cuéllar. "Government by Algorithm: Artificial Intelligence in Federal Administrative Agencies." Artificial Intelligence in Federal Agencies , February 2020. https://www-cdn.law.stanford.edu/wp-content/uploads/2020/02/ACUS-AI-Report.pdf.

"Estonia - We Have Built a Digital Society and We Can Show You How." e-estonia, February 25, 2021. https://e-estonia.com/.

"The Film." simorgh, n.d. https://www.simorghanimation.com/the-film.

Finnegan, Patrick. *Wind Turbine Blades.* June 28, 2008. *Flickr.* https://flickr.com/photos/vax-o-matic/2621890270.

Fried, Morton H. *The Evolution of Political Society: an Essay in Political Anthropology.* New York: Random House, 1967.

Frye, R N. "Chapter 4: The Political History of Iran under the Sasanians." Essay. In *The Cambridge History of Iran* 3, Vol. 3. Cambridge: Univ. Press, 1983.

Gannon, By Megan. "Ancient Copy of Ten Commandments Goes Digital." NBCNews.com. NBCUniversal News Group, December 13, 2012. https://www.nbcnews.com/id/wbna50181707.

Gerard, R. W., Russell B. Stevens, Charles Molnar, and Jane Gair. "Chapter 7: Introduction to the Cellular Basis of Inheritance." Essay. In *Concepts of Biology*. Washington, D.C., VA: National Academy of Sciences, 1958.

Gonzalez, Carlos. "What's the Future Role for Humanoid Robots?" MachineDesign, October 27, 2017. https://www.machinedesign.com/mechanical-motion-systems/article/21836113/whats-the-future-role-for-humanoid-robots.

Goswamy, B N. "What the Bird Prophesised." The Tribune, November 25, 2018. https://www.tribuneindia.com/news/archive/features/what-the-bird-prophesised-687308.

"Great Educator: Baron De Montesquieu 1689 To 1755." Ragged University, November 24, 2016. https://www.raggeduniversity.co.uk/2011/10/05/montesquieu-and-cultural-relativism-1689-1755/.

Hamilton, Alexander. "The Federalist Papers: No. 68." The Avalon Project : Federalist No 68. Accessed July 12, 2021. https://avalon.law.yale.edu/18th_century/fed68.asp.

Harris, William V. *Ancient Literacy*. New York, NY: ACLS History E-Book Project, 2004.

Herculano-Houzel, S., Mariano, L. "A Comparison of Encephalization between Odontocete Cetaceans and Anthropoid Primates" Brain Behavior and Evolution No. 51, pp 230-238, 1998.

History.com Editors. "Code of Hammurabi." History.com. A&E Television Networks, November 9, 2009. https://www.

history.com/topics/ancient-history/hammurabi.

Hoffman, Donald D. "Sensory Experiences as Cryptic Symbols of a Multimodal User Interface." *Activitas Nervosa Superior* 52, no. 3-4 (2010): 95–104. https://doi.org/10.1007/bf03379572.

Hollister, Sean. "At Last -- Smart Glasses That Don't Look like Borg Headgear." CNET, February 5, 2018. https://www.cnet.com/tech/mobile/intel-vaunt-smart-glasses-prototype/.

IEEE Spectrum, May 2016. https://www.nxtbook.com/nxtbooks/ieee/spectrum_na_0516/index.php#/p/Cover1.

"Invention of the Television." DK Find Out!, n.d. https://www.dkfindout.com/us/science/amazing-inventions/television/.

"Iran-Contra Affair." Encyclopædia Britannica. Encyclopædia Britannica, inc., n.d. https://www.britannica.com/event/Iran-Contra-Affair.

Jaganmohan, Madhumitha. "Energy Consumption of Alphabet (Google) from Financial Year 2011 to 2019." Statista, March 24, 2021. https://www.statista.com/statistics/788540/energy-consumption-of-google/.

Kardashev, N. S. "Transmission of Information by Extraterrestrial Civilizations." *Soviet Astronomy* 8 (1964): 217–21.

"Keypunch." Wikipedia. Wikimedia Foundation, March 31, 2021. https://en.wikipedia.org/wiki/Keypunch.

Kierkegaard, Søren. *Philosophical Fragments; Johannes Climacus.* Translated by Howard Vincent Hong. Princeton, NJ: Princeton University Press, 1987.

"Killer Facts 2019: The Scale of the Global Arms Trade." Killer facts: The scale of the global arms trade | Amnesty International, August 23, 2019. https://www.amnesty.org/en/latest/news/2019/08/killer-facts-2019-the-scale-of-the-global-arms-trade/.

Kochenderfer, Wheeler, and Wray, *Algorithms for Decision Making,* MIT Press, 2022.

Lampinen, James M., Jeffrey S. Neuschatz, and David G. Payne. "Memory Illusions and Consciousness: Examining the Phenomenology of True and False Memories." *Current Psychology* 16, no. 3-4 (1997): 181–224. https://doi.org/10.1007/s12144-997-1000-5.

Levin, Michael. "What Bodies Think About: Bioelectric Computation Outside the Nervous System, Primitive Cognition, and Synthetic Morphology." NeurIPS, December 4, 2018. https://neurips.cc/Conferences/2018/Schedule?showEvent=12487.

"Library Catalog." Library catalog - New World Encyclopedia, n.d. https://www.newworldencyclopedia.org/entry/Library_catalog.

Marx, Karl. *Critique of the Gotha Program*, 1875.

Mathilde Thomas, *The French Beauty Solution: Time-Tested Secrets to Look and Feel Beautiful Inside and Out*, Penguin, Jul 14, 2015.

McNamara, Patrick. "Why Some of Your Dreams Have Sequels." Psychology Today, December 30, 2014. https://www.psychologytoday.com/us/blog/dream-catcher/201412/why-some-your-dreams-have-sequels.

Mikkelson, David. "Is This a Real Photograph of an 'Insect Spy Drone?'" Snopes.com, November 17, 2020. https://www.snopes.com/fact-check/insect-spy-drone/.

Mohamed, Alana. "How J. Edgar Hoover Used the Power of Libraries for Evil." Literary Hub, March 5, 2020. https://lithub.com/how-j-edgar-hoover-used-the-power-of-libraries-for-evil/.

Monte, Jonas. "Sum, Ergo Cogito: Nietzsche Re-Orders Descartes." *Aporia* 25, no. 2 (2015): 13–23.

Murano, Grace. "12 Most Amazing Bird Formations - Murmuration, Amazing Birds." Oddee, September 12, 2012. https://www.oddee.com/item_98319.aspx.

Nebbia, Giorgio, and Gabriella Nebbia Menozzi. *A Short History of Water Desalination*. Milan: Azienda Grafica Italiana, 1966.

Netburn, Deborah. "430,000-Year-Old Skull Suggests Murder Is an 'Ancient Human Behavior.'" *Los Angeles Times*, May 29, 2015. https://www.latimes.com/science/sciencenow/la-sci-sn-earliest-known-murder-victim-20150526-story.html.

"Neurons with Thousands of Connections: Where Are the Extra Connections Coming from?" Biology Stack Exchange, March 1, 1963. https://biology.stackexchange.com/questions/22011/neurons-with-thousands-of-connections-where-are-the-extra-connections-coming-fr.

Nicol, Will. "The 9 Coolest Military ROBOTS: Maars, DOGO, Etc." Digital Trends. Digital Trends, July 17, 2020. https://www.digitaltrends.com/cool-tech/coolest-military-robots/.

"Nicole King." Nicole King | Research UC Berkeley, September 1, 2017. https://vcresearch.berkeley.edu/faculty/nicole-king.

"Optische Illusie - Over Fysiologische Illusies En Schijnwaarnemingen." Pieter Broertjes, January 20, 2020. https://www.pieterbroertjes.nl/optische-illusie/.

"Other Services (except Public Administration) in the US." IBISWorld. Accessed August 19, 2021. https://www.ibisworld.com/industry-trends/market-research-reports/other-services-except-public-administration/repair-maintenance/.

"Pascal's Calculator." Wikipedia. Wikimedia Foundation, March 9, 2021. https://en.wikipedia.org/wiki/Pascal%27s_calculator.

Patel, Bhavesh H., Claudia Percivalle, Dougal J. Ritson, Colm D. Duffy, and John D. Sutherland. "Common Origins of RNA, Protein and Lipid Precursors in a Cyanosulfidic Protometabolism." *Nature Chemistry* 7, no. 4 (2015): 301–7. https://doi.org/10.1038/nchem.2202.

Percival, Allen. *A History of Music*. The English Universities, 1961.

Pollard, Georgia, James Ward, and Philip Roetman. "Water Use Efficiency in Urban Food Gardens: Insights from a Systematic Review and Case Study." *Horticulturae* 4, no. 3 (September 12, 2018): 27. https://doi.org/10.3390/horticulturae4030027.

Powers, Richard Gid. *Secrecy and Power the Life of J. Edgar Hoover.* London: Arrow Books, 1989.

Price, Rickey. "Please Stand by: A Journey through the History of Television Super Heroes." Comic Watch, November 23, 2019. https://comic-watch.com/news/please-stand-by-a-journey-through-the-history-of-television-super-heroes.

Rada, Camilo. "What Is the Difference between Radiation Balance and the Global Energy Balance?" Earth Science Stack Exchange, March 11, 2019. https://earthscience.stackexchange.com/questions/13600/what-is-the-difference-between-radiation-balance-and-the-global-energy-balance.

Resnick, Brian. "Can California Make It Rain with Drones?" The Atlantic, October 7, 2014. https://www.theatlantic.com/politics/archive/2014/10/can-california-make-it-rain-with-drones/453393/.

Ritchie, Hannah, and Max Roser. "Meat and Dairy Production." Our World in Data, August 25, 2017. https://ourworldindata.org/meat-production.

Rivest, R. L., A. Shamir, and L. Adleman. "A Method for Obtaining Digital Signatures and Public-Key Cryptosystems." *Communications of the ACM* 21, no. 2 (February 1978): 120–26. https://doi.org/10.1145/359340.359342.

"Robotic Exoskeleton, the External Skeleton Made for Humans." Ozwana. Accessed July 27, 2021. http://ozwana.blogspot.com/2014/03/robotic-exoskeleton-external-skeleton.html.

Roser, Max. "Employment in Agriculture." Our World in Data, April 26, 2013. https://ourworldindata.org/employment-in-agriculture.

Roser, Max. "Employment in Agriculture." Our World in Data, April 26, 2013. https://ourworldindata.org/employment-in-agriculture.

Sapkota, Mukta, Meenakshi Arora, Hector Malano, Magnus Moglia, Ashok Sharma, Biju George, and Francis Pamminger. "An

Overview of Hybrid Water Supply Systems in the Context of Urban Water Management: Challenges and Opportunities." *Water* 7, no. 12 (December 29, 2014): 153–74. https://doi.org/10.3390/w7010153.

Scheve, Tom. "How Body Farms Work." HowStuffWorks Science. HowStuffWorks, January 27, 2020. https://science.howstuffworks.com/body-farm1.htm.

"The Science of Emotion: Exploring the Basics of Emotional Psychology." UWA Online, June 27, 2020. https://online.uwa.edu/news/emotional-psychology/.

Scutti, Susan. "What It's Like To Experience Synesthesia: The Taste Of Music And Colors Of Language." Medical Daily, March 7, 2014. https://www.medicaldaily.com/what-its-experience-synesthesia-taste-music-and-colors-language-270741.

Secondat, Montesquieu Charles de. *The Spirit of Laws Translated from the French of M. De Secondat, Baron De Montesquieu. A New Translation. In Three Volumes. ..* Berwick: Printed for R. Taylor, 1770.

"Semi-Autonomous & Autonomous Truck Market." Market Research Firm, May 2020. https://www.marketsandmarkets.com/Market-Reports/semi-autonomous-truck-market-224614273.html.

Shamiri, Cyrus. "Nashtifan, the Ancient City of Windmills." History Forum, June 10, 2009. http://archive.worldhistoria.com/nashtifan-the-ancient-city-of-windmills_topic27376.html.

Shuttleworth, Jennifer. "SAE J3016 Automated-Driving Graphic." SAE International, January 1, 2019. https://www.sae.org/news/2019/01/sae-updates-j3016-automated-driving-graphic.

Simon, Matt. "Embodied Intelligence Wants to Teach Robots with Virtual Reality." Wired, November 7, 2017. https://www.wired.com/story/embodied-intelligence-want-to-really-teach-a-robot-command-it-with-vr/.

"Simple Transposition Ciphers." Crypto Corner, n.d. https://crypto. interactive-maths.com/simple-transposition-ciphers.html.

Sindico, Francesco, Ricardo Hirata, and Alberto Manganelli. "The Guarani Aquifer System: From a Beacon of Hope to a Question Mark in the Governance of TRANSBOUNDARY AQUIFERS." *Journal of Hydrology: Regional Studies* 20 (2018): 49–59. https://doi.org/10.1016/j.ejrh.2018.04.008.

"The Solar Constant." Australian Government - Bureau of Meteorology, Space Weather Services. Accessed July 27, 2021. https:// www.sws.bom.gov.au/Educational/2/1/12#:~:text=The%20 luminosity%20of%20the%20Sun,and%20X%2Dray%20 spectral%20bands.

"Stanford University Research Found That Parents Must Adhere to These 4 Things If They Want to Promote Their Children's Brain Development." inf.news, August 31, 2021. https://inf. news/en/baby/329e4626035136f8f64e52e5ce0f1bdf.html.

"TET 2018 - Chapter 4 - Transportation Employment." TET 2018 - Chapter 4 - Transportation Employment | Bureau of Transportation Statistics, February 14, 2021. https://www. bts.gov/transportation-economic-trends/tet-2018-chapter-4-employment.

"Together, We Will Find the Cure!" The Progeria Research Foundation, September 2, 2021. https://www.progeriaresearch.org/.

Tran, Pheobe. "Amazon Go to Open Six More Locations, ALBERTSONS Buys Rite Aid as Amazon Threat LOOMS + More." Food+Tech Connect, April 10, 2018. https:// foodtechconnect.com/2018/02/22/amazon-go-to-open-six-more-locations-albertsons-buys-rite-aid-as-amazon-threat-looms-more/.

Tzu, Sun. *The Art of War*, n.d.

U.S. Department of Transportation, and Federal Aviation Administration. *A Plan for the Future: 2006-2015: The Federal Aviation ADMINISTRATION'S 10 Year Strategy for the Air Traffic CONTROL Workforce*. Washington, D.C., VA: Federal

Aviation Administration, 2016.

Vaheddoost, Babak, Javad Behmanesh, Hafzullah Aksoy, and Hossien Rezaie. "Estimating the Effect of Qanats and Underground Dam on Water Levels in Wells, Using Finite Difference Simulation." *ResearchGate*, May 2014. https://doi.org/10.13140/RG.2.1.1117.9684.

Walker, M. B. *German Bayonetting Children*. July 25, 1915. *Life*.

"Water Resources, China: Short-Term Fixes." Law-In-Action, October 8, 2014. https://law-in-action.com/2014/10/08/the-price-for-self-sufficiency-water-resources-in-china/.

Whitehead, Alfred North, and Bertrand Russell. *Principia Mathematica*. Cambridge, MA: Cambridge University Press, 1910.

Wolchover, Natalie. "Cows Make Humanized Milk. But Is It Safe?" LiveScience. Purch, June 10, 2011. https://www.livescience.com/14538-cows-humanized-milk-safe.html.

Yaffe, Martin D., and Richard S. Ruderman. *Reorientation: Leo Strauss in the 1930s*. New York, NY: Palgrave Macmillan, 2014.

Zacevini, Gianni. "Il Primo Telefono a Milano." Divina Milano, June 3, 2021. https://www.divinamilano.it/il-primo-telefono-a-milano/.

Majid Riaziat is an experienced scientist and entrepreneur with continued involvement in technology development and commercialization. Dr. Riaziat received his Ph.D. in Applied Physics from Stanford University. His employment background includes Varian Medical Systems, where he was a research director, and OEpic Semiconductors, where he was cofounder and CEO.